© 1998 by Eli Steele

SHELBY STEELE is a Research Fellow at the Hoover Institution, Stanford University. His work has appeared in *Dissent*, the *New Republic*, *Harper's*, *Newsweek*, the *Wilson Quarterly*, the *New York Times*, the *Los Angeles Times*, the *Washington Post*, the *Wall Street Journal*, *Commentary*, and *American Scholar*, among other publications. *The Content of Our Character* won the National Book Critics Circle Award for nonfiction in 1991.

A DREAM DEFERRED

THE SECOND BETRAYAL OF BLACK FREEDOM IN AMERICA

SHELBY STEELE

HarperPerennial
A Division of HarperCollins*Publishers*

"The New Sovereignty" appeared in *Harper's Magazine* in June 1992.

Grateful acknowledgment is made for permission to reprint
"Dream Deferred": From *Collected Poems* by Langston Hughes.
Copyright © 1994 by the Estate of Langston Hughes.
Reprinted by permission of Alfred A. Knopf, Inc.

A hardcover edition of this book was published in 1998 by
HarperCollins Publishers.

HarperCollins books may be purchased for educational, business, or sales
promotional use. For information please write: Special Markets Department,
HarperCollins Publishers, Inc., 10 East 53rd Street, New York, NY 10022.

First HarperPerennial edition published 1999.

Designed by Kyoko Watanabe

The Library of Congress has catalogued the hardcover edition as follows:

Steele, Shelby.
A dream deferred : the second betrayal of black freedom in America /
Shelby Steele.—1st ed.
p.   cm.
Contents: Preface—The loneliness of the "black conservative"—
Wrestling with stigma—Liberal bias and the zone of decency—
The new sovereignty.
ISBN 0-06-016823-4
1. United States—Race relations.   2. Afro-Americans—Civil rights—
History—20th century.   3. Afro-Americans—Politics and government.
4. Conservatism—United States—History—20th century.
I. Title.
E185.615.S7234 1998
305.8'00973—dc21   98-18483

ISBN 0-06-093104-3 (pbk.)

04 05 ❖/RRD 10 9 8 7 6 5

For Rita

But there is also an American Negro tradition which teaches one to deflect racial provocation and to master and contain pain. It is a tradition which abhors as obscene any trading on one's anguish for gain or sympathy; which springs not from a desire to deny the harshness of existence but from a will to deal with it as men at their best have always done.

RALPH ELLISON,
Shadow and Act

Try to remember that human dignity is an absolute, not a piece-meal notion; that it is inconsistent with special pleading. . . . Should you find this argument a bit on the heady side, think at least that by considering yourself a victim you but enlarge the vacuum of irresponsibility that demons or demagogues love so much to fill. . . .

Maybe the real civility, Mr. President, is not to create illusions.

JOSEPH BRODSKY,
On Grief and Reason

What happens to a dream deferred?

Does it dry up
like a raisin in the sun?
Or fester like a sore—
And then run?

Does it stink like rotten meat?
Or crust and sugar over—
like a syrupy sweet?
Maybe it just sags
like a heavy load.

Or does it explode?

LANGSTON HUGHES,
"Dream Deferred"

CONTENTS

PREFACE

If there is an insight that unifies the four essays that comprise this book, it is that America's collision with its own racial shame in the civil rights era is the untold story behind today's polarized racial politics. A society is very dangerous to itself when it has to bear an undeniable shame. There will be a powerful impulse to redeem itself by betraying its best principles, to bend and suspend those principles in order to show that its remorse over its shame is deeper than any priggish commitment to great principles. In other words, self-betrayal can become the road to redemption for the shamed society. Self-betrayal can form the new redeeming idea of social virtue; it can become the basis of a new redemptive politics.

These essays contend that the liberalism that grew out of the sixties was such a politics—that its first and all-consuming goal was the expiation of American shame rather than the careful and true development of equality between the races. Shame pushed the post-sixties United States into an extravagant, autocratic, socialistic, and interventionist liberalism that often betrayed America's best principles in order to give whites and American institutions an iconography of racial virtue they could use against the stigma of racial shame. An implication of this work is that our ceaseless debate over affirmative action is, in fact, a debate over the peculiar liberalism generated by shame.

My highest hope for these essays is that they will be explanatory. In some places the reader may notice what I hope is a tolerable amount of repetition. One reason for this is that the book consists of three essays ("The New Sovereignty" was published earlier), each of which needed to stand on its own terms. This necessitated some repetition of background points developed in other essays. Another reason is that my experience of writing about America's racial conundrum is not unlike that of poor Sisyphus, who was forever bracing himself for yet another trudge up the same mountain. Though the foothills and high ridges may repeat, like him, I feel that this is all I can do, though I hope always for new meaning.

Another difficulty in writing about race is that a national discussion of it is always raging on while one is trying to figure things out. This can foster the feeling that one is chasing an ever-changing phenomenon, an elusive animal that escapes every grip. But I have tried to remember that this is an illusion. When our racial story changes, it usually does so in familiar ways. With race there are more new events to consider than fundamental changes. And in this work I have tried not to hear so much noise that I forget that it is old patterns that need understanding. There is a lot of déjà vu in this sort of work, and that is as it must be.

I would like to thank the Hoover Institution at Stanford University for its support during the writing of this book. The rule of thumb at Hoover is that its scholars have opinions but the Hoover Institution does not. I cannot imagine a more congenial and inspiring place to work, and I am indeed grateful for the association.

I would also like to thank my editor at HarperCollins, Terry Karten. She stayed with this book above and beyond the call of duty, so that her belief in it helped it to materialize. I thank my agent, Carol Mann, for her tireless faith and support.

Last, I want to thank my wife, Rita, daughter, Loni, and son, Eli, for the many hours they spent carefully reading over the various drafts of these essays. I also thank Loni and Eli for the title of this book and for their insight that it conveys my truest meaning.

THE LONELINESS OF THE "BLACK CONSERVATIVE"

1.

I felt a familiar anger rise when the editor asked me over the phone to write about "the loneliness of a black conservative." Unless this was to be one of those serendipitous matchups of writer and subject where the charm is in the incongruence, the request all but stated that I was an obvious choice to map out this new territory of American loneliness. But the anger I felt was immediately diffused by an equally familiar sense of fatalism. There was no point in arguing. To be called a black conservative is, in fact, to be one, or at least to pay the price for being one. Besides, my life has been varied enough that I can now lay reasonable claim to many black identities, black conservative among them. As for loneliness, it is no doubt a risk that trails every effort to define one's beliefs. Most people could empty half of any room simply by saying what they truly believe. If, somehow, you come by the black conservative imprimatur, you will likely empty a lot more than half the room *before* you say what you believe.

I realized, finally, that I was a black conservative when I found myself standing on stages being shamed in public. I had written a book that said, among many other things, that black American leaders were practicing a politics that drew the group into a victim-focused racial identity that, in turn, stifled black advancement more than racism itself did. For reasons that I will discuss shortly, this was heresy in many quarters. And, as I traveled around from one little Puritan village (read "university") to another, a common scene would unfold.

Whenever my talk was finished, though sometimes before, a virtual militia of angry black students would rush to the microphones and begin to scream. At first I thought of them as Mau Maus, but decided this was unfair to the real Mau Maus, who, though ruthless terrorists, had helped bring independence to Kenya in the 1950s. My confronters were not freedom fighters; they were Carrie Nation–like enforcers, racial bluenoses, who lived in terror of certain words. Repression was their game, not liberation, and they said as much. "You can't *say* that in front of the white man." "Your words will be used against us." "Why did you write this book?" "You should only print that in a black magazine." Their outrage brought to light an ironic and unnoticed transformation in the nature of black American anger from the sixties to the nineties: a shift in focus from protest to suppression, from blowing the lid off to tightening it down. And, short of terrorism, shame is the best instrument of repression.

Of course *most* black students did not behave in this way. But the very decency of the majority, black and white, often made the shaming of the minority more effective. So I learned what it was like to stand before a crowd in which a coterie of one's enemies had the license to shame, while a mixture of decorum and fear silenced the decent people who might have come to one's aid. I was as vulnerable to the decency as to the shaming, since together they amounted to shame. And it is never fun to be called "an opportunist," "a house slave," and so on while university presidents sit in the front row and avert their eyes. But this really is the point: The goal of shaming was never to win an argument with me; it was to make a *display* of shame that would make *others* afraid for themselves, that would cause eyes to avert. I was more the vehicle than the object, and what I did was almost irrelevant. Shame's victory was in the averted eyes, the cowering of decency.

Today a public "black conservative" will surely meet a stunning amount of animus, demonization, misunderstanding, and flat-out, undifferentiated contempt. And there is a kind of licensing process involved here in which the black leadership—normally protective even of people like Marion Berry and O. J. Simpson—licenses blacks *and* whites to have contempt for the black conservative. It is a part of the group's manipulation of shame to let certain of its members languish outside the perimeter of group protection where even politically correct whites (who normally repress criticism of blacks) can show contempt for them.

Not long ago I heard a white female professional at a racially mixed dinner table call Clarence Thomas an incompetent beneficiary of affirmative action—the same woman whom I had heard on another occasion sneer at the idea that affirmative action stigmatized women and minorities as incompetent. Feminists who happily vote for Bill Clinton are free to loathe Clarence Thomas. In a sense Thomas Sowell, Walter Williams, Ward Connerly, Stanley Crouch, myself, and many, many others represent a new class of "unprotected" blacks. By my lights there is something a little avant-garde in this. But, as with any avant-garde, the greater freedom is paid for in a greater exposure to contempt and shame.

The Czech writer Milan Kundera—a man whose experience under the hegemony of the Communist Party taught him much about the shaming power of groups over the individual—says that shame transforms a person "from a subject to an object," causes them to lose their "status as individuals."

And to suffer this fate means that the group—at least symbolically—has determined to annihilate you. Of course we have no gulags in black America, but black group authority—like any group authority—defines itself as much by who it annihilates as by who it celebrates. Thus it not only defines group, it also defines *grouplessness*. And here, on this negative terrain, where his or her exclusion sharpens the group identity, the black conservative lingers as a kind of antithesis.

But is this loneliness? I'm not sure.

The problem for the black conservative is more his separation from the *authority* of his racial group than from the actual group. He stands outside a group authority so sharply defined and monolithic that it routinely delivers more than 90 percent of the black vote to *whatever* Democrat runs for president. The black conservative (who for convenience I will sometimes abbreviate down to "BC") may console himself with the idea that he is on the side of truth, but even truth is cold comfort against group authority (which very often has no special regard for truth). White supremacy focused white America's group authority for three centuries before truth could even begin to catch up. Group authority is just as likely to be an expression of collective ignorance as of truth; but it is always, in a given era, more powerful than truth.

This authority is very often based on a strategic explanation of a group's fate, a narrative that explains why the group is in a given situation and therefore why it is justified in pursuing a certain kind of power. This explanation is all-important because it establishes the group as a collective being with a history, a present, and a future—a life, as it were, that entitles it to all the considerations of sovereignty. In the schools of every nation, children hear the *story* of their country's struggle for sovereignty. But for a minority group like American blacks, whom history has left with a deep sense of vulnerability, shame becomes a primary means of reinforcing the group's story. Shame provides the muscle to keep individuals in line with group authority.

And shame does this muscling by making conformity to the group explanation the measure of one's love for the group. Thus nonconformity is a failure of love, a betrayal. And this is the most constant charge against the black conservative—that he does not love his own people—an unpardonable sin that justifies his symbolic annihilation. Because the capacity to love makes us human, it is precisely the charge that a person is without love that transforms him "from subject to object" and causes him to lose his "status as individual(s)." So nonconformity not only points to a failure of love but also to a kind of inhumanity.

All of this is made worse by the fact that black Americans have been a *despised* minority surrounded by indifference and open hatred. An individual's failure of group love is a far greater infraction among blacks because it virtually allies that individual with the enemy all around. An Uncle Tom is someone whose failure to love his own people makes him an accessory to their oppression. So group love (in one form or another) is a preoccupation in black life because of the protective function it serves, because we want to use the matter of love as a weapon of shame, and thus as an enforcer of conformity. Love adds the seriousness and risk to nonconformity.

If this gives black America the means to enforce its group authority—and its explanation of its fate—it also plagues us with a repressive, one-party politics. Because of historic vulnerability and the resulting insistence on conformity around a single strategic explanation of group fate, black America has not yet achieved a two-party politics. Thus black conservatives do not yet comprise a loyal opposition; they are, instead, classic dissenters. This differentiates them from white conservatives, who work out of a two-party group. In his dissent from a one-party-one-explanation group politics, the BC lives the life of a dissenter, a life too conspicuously gambled on belief, a life openly subversive to his own group and often impractical for himself—a life at odds.

What, in fact, is a black conservative?

Well, he is not necessarily a Republican, or free-market libertarian, or religious fundamentalist, pro-lifer, trickle-down economist, or neocon. I have met blacks in all these categories who are not considered conservatives.

The liberal-conservative axis is a bit different for blacks than for Americans generally. Under his American identity a black Republican is conservative, but under his racial identity he may be quite liberal. Many black Republicans, for example, are intense supporters of preferential affirmative action and thus liberal in terms of their group identity. (Colin Powell is a case in point, as is Arthur Fletcher, a black Republican who helped President Nixon introduce America's first racial preference in the famous "Philadelphia Plan.") But the "new" black conservatives—the ones who have recently become so controversial—may even be liberal by their American identity but are definitely conservative by the terms of their group identity. It is their dissent from the *explanation* of black group authority that brings them the "black conservative" imprimatur. Without this dissent we may have a black Republican but not a "black conservative," as the term has come to be used.

And what is this explanation? In a word it is victimization. Not only is victimization made to explain the hard fate of blacks in American history, but it is also asked to explain the current inequalities between blacks and whites and the difficulties blacks have in overcoming them. Certainly no explanation of black difficulties would be remotely accurate were it to ignore racial victimization. On the other hand, victimization does not in fact explain the entire fate of blacks in America, nor does it entirely explain their difficulties today. It was also imagination, courage, the exercise of free will, and a very definite genius that enabled blacks not only to survive victimization but also to create a great literature, utterly transform Western music, help shape the American language, expand and deepen the world's concept of democracy, influence popular culture around the globe, and so on. No people with this kind of talent, ingenuity, and self-inventiveness would allow victimization so singularly to explain their fate unless it had become a primary source of power. And this is precisely what happened after the sixties. Victimization became so rich a vein of black power—even if it was only the power to "extract" reforms (with their illusion of deliverance) from the larger society—that it was allowed not only to explain black fate but to explain it *totally*.

A black woman journalist I met recently for lunch said: "I don't think we can tell the story of our victimization enough." We were talking about an article she was writing. She was young, Ivy League–educated, and, sitting across from me in the patio restaurant, she might have been an advertisement for any number of blessings—good health, good upbringing, good fortune. Politely we argued about how much victimization blacks were still subjected to. I said it was number three or four on the list of things that held blacks back. She said it was number one. And here we had arrived at one of the most telling impasses two black Americans can reach. Her number one ranking aligned her with the explanation of black fate on which black group authority rests. For her, victimization was not a fact of black life, it was *the* fact. It was a totalism—an ultratruth that not only supersedes but that makes a *taboo* of all other truths. My lower ranking of racism as a barrier violated this taboo, put me at odds with black group authority, and made me, alas, a "black conservative."

Very simply, then, a black conservative is a black who dissents from the victimization explanation of black fate *when it is offered as a totalism*—when it is made the main theme of group identity and the raison d'être of a group politics.

The young journalist was a liberal and in harmony with black group authority because of a predetermined willingness, even commitment, to seeing two things: that black difficulty in America was the result of ongoing racial victimization and that white America was responsible for bringing change. The only time she transgressed her natural politeness was when she smugly said, "Well, obviously we have a different time schedule as to when white people ought to be let off the hook." Certainly even a BC would not want to let white people off the hook. And yet, as time marches on, I can't help but feel that a far greater danger for blacks is the belief that doing so makes a difference. What is clear is that a group politics devoted to keeping whites on the hook also requires that victimization be a totalism in black life—that it define group identity, become a part of the self-image of individual blacks, and keep in play a permanently contentious relationship with whites.

I said to her that when victimization is treated as a totalism, it keeps us from understanding the true nature of our suffering. It leads us to believe that *all* suffering is victimization and that *all* relief comes from the guilty good-heartedness of others. But people can suffer from bad ideas, from ignorance, fear, a poor assessment of reality, and from a politics that commits them to the idea of themselves as victims, among other things. When black group authority covers up these other causes of suffering just so whites will feel more responsible—and stay on the hook—then that authority actually encourages helplessness in its own people so that they might be helped by whites. It tries to make black weakness profitable by selling it as the white man's burden.

"But isn't it really about power? And if victimization brings power, it's the power that counts." She surprised me. I hadn't realized she was even listening. "I mean, you could say that whites got power by killing the Indians and enslaving the blacks. That's worse than using your history of victimization to get power. People get power all sorts of ways."

We were outside the restaurant now and she was hurrying to cover the few blocks to her rented car to make her next appointment. Working to keep pace, I suddenly felt a familiar doubt. Let's call it the black conservative doubt—the feeling that one is talking into a void, that one might be right, might even have a compelling piece of truth, but that it is a truth unattached to any necessity, a truth with no means of enforcing itself. Often people don't listen as much for the truth as for the necessity that will hold them accountable to the truth. Failing to hear any such necessity, they can conclude that the truth itself has no relevance.

The great problem for the BC is that the necessity of his or her truth is hidden so that it seems irrelevant, academic. What keeps it hidden is the symbiosis between whites and blacks by which they agree to let victimization *totally* explain black difficulty. Whites agree to stay on this hook for an illusion of redemption, and blacks agree to keep them there for an illusion of power. I can say that these investments are illusions, that whites have no real redemption and that blacks have no real power, but then what do I have? That's really what the young journalist was saying to me as we walked to her car. Government, corporate America, universities, foundations—they were all in the business of seeing blacks as victims, of trading an illusion of power for an illusion of redemption. Everybody was practiced in these negotiations, so the fact that they encouraged helplessness in blacks, kept them mired in a victim-focused identity, gave them a disinvestment in success and an investment in failure . . . well. The BC is at odds with a very cozy and very functional symbiosis, and there is always something to be said for function. He may believe that there are bodies under the floorboards, but until that truth is more widely understood, there is not much necessity in what he says.

I was not surprised when we turned a corner and came upon the journalist's rental car. It was a huge, white Lincoln Towncar with plush leather upholstery, and it sat so regally on the street that the smaller cars around it seemed to compose its court. I thought she might apologize for it, as people often do with their ostentations, but she said only that she had a "good" expense account. After quickly shaking my hand good-bye, she swung open the driver's door and all but plunged in. In a moment the white boat was floating down the street.

2.

After the sixties, when American politics became openly accountable to the legacy of racial victimization, the acceptance or rejection of victimization as a totalism came to imply either a liberal or conservative politics. In response to the sixties American liberalism realigned itself around victimization not as a fact or as an ongoing problem but as a totalistic explanation of black difficulty. Conservatism during this period belatedly admitted to the fact of black victimization but never accepted it as a totalism. To a profound degree this relation to the totalism of victimization came to demarcate social liberalism and conservatism after the sixties. And to this day the liberal looks at black difficulties—high crime rates, weak academic performance, illegitimacy rates, and so on—and presumes them to be the result of victimizing forces beyond the control of blacks. The conservative does not deny this as a possibility but refuses to *presume* it. This refusal has become a contemporary mark of social conservatism.

I believe that this acceptance of victimization as a totalism caused the downfall of post-sixties liberalism. This is where liberalism lost its balance and ultimately its integrity. Many observers who lived through the sixties realize that it was the old American problem of race that did liberalism in. To accept victimization not as one of many variables but as a totalism was to see it as *structural*—so built into the patterns of society that it could be manifested apart from human will. And if the evil was structural, only structural remedies would work against it. You couldn't fight racial victimization on a case-by-case basis; you had to put into place structures that would prefer the victim in compensation for the victimization we could *presume* he or she had endured. Thus liberalism became *preemptive* rather than defensive. It no longer protected individuals and fought for equal opportunity but it pursued group rights and equal results. It remedied the victimization before it was manifest. This transformation came from the embrace of victimization as a totalistic explanation of black difficulty. But it changed the basic terms of American liberalism from freedom, rights, and responsibilities to planning, engineering, and entitlements.

And so the black conservative was created by default. Liberalism moved away from him and into socialism, so that his "conservatism" resulted as much from his being abandoned as from anything else. Today a liberal does not believe in or work for a color-blind society; that is what conservatives want. Today a liberal supports the government's right to make distinctions by race (for purposes of social engineering) just as segregationists fought for the same right thirty-five years ago. Today racial and ethnic groups are celebrated by liberals while conservatives celebrate the individual with quotations from Martin Luther King, Jr.

The black conservative is unable to swallow the one idea that would have enabled him to follow liberalism in its preemptive and socialistic turn: racial victimization as a totalism. And so, failing to swallow, he also fails to join. But does he feel alone?

The black conservative not only dissents from this totalism; He or she is also at odds with America's idea of how to remedy the damage of historical victimization. This idea can broadly be referred to as *structuralism*.

The black conservative's problem with structuralism is what I like to refer to as the Bigger Thomas problem.

In what is certainly one of the most "structural" incidents in American literature, Bigger Thomas—the black protagonist of Richard Wright's novel *Native Son*—presses a pillow into the white face of young Mary Dalton until she is dead. Bigger had not meant to kill her. He had just been hired earlier that same day as the Dalton family chauffeur, and after driving Mary and her boyfriend around Chicago for a night of carousing, he found that Mary was so drunk she could not make it inside to bed. This is the late 1930s, and Bigger is a poor, young, uneducated black male living a life so constricted—and, in an odd way, protected—by racial oppression that he experiences whites not as people but as awesome natural forces that fill him with terror. And yet beneath this terror there is a sullen and simple lust that registers as he carries the drunken Mary upstairs to bed. As he pauses over her, no more than watching her, he suddenly hears the voice of her mother calling her name from the hallway. He freezes in abject and mindless terror. He and the entire world know what it means for someone like him to be caught in the bedroom of someone like her in the middle of the night. Without any thought at all, in an act of hysterical self-preservation, he presses the pillow into her face so that she cannot answer her mother's soft call. Her fingers dig into his arm before she goes still. The mother hears nothing and leaves. Mary Dalton is dead.

One of the great ironies in this scene stems from the fact that Mary's mother is blind. Bigger could simply have stood there and gone undetected. He could quietly have slipped out of the room or lingered in a corner until the mother left. Or better yet, he could have told the mother directly that he had to put her drunken daughter to bed and that he would appreciate better behavior from her in the future. But fear animalized Bigger, cut him off from his own reason, and drove him to play out the worst of all possible options.

Where did the fear come from? It came from the racism of white Americans that, over time, had congealed into intractable social structures designed to keep the races separate. And the greatest structural taboo of all was against sex between the black male and the white female. The mere charge of sexual interest in a white woman had brought death to many a black man. Richard Wright was a social determinist, and his goal was to show the determining power of racist social structures. Thus Bigger was so determined by these structures, so infused with and terrorized by the taboo they carried, that it was society, not him, that killed Mary Dalton. It was the hateful racial terrorism of society that pressed the pillow into her face. And, paradoxically, there was an innocence in this murderer whose killing was only an outgrowth of his victimization.

The great insight of structuralism was that an evil like

racism could have an impersonal life in the structures of society (customs, manners, residential living patterns, folkways, even laws in the days of segregation). It could determine events apart from the will of individuals or even groups. Mary Dalton had liked Bigger, and he in turn had had no grudge against her. The event of her death was an impersonal playing out of structural forces—the "invisible hand" of racism. People could understand that Bigger Thomas, inadvertent murderer that he was, was also an American creation. So structuralism is really an idea about the relationship between the social attitude and the event, between the prejudice and the oppression. America needed this idea to understand that segregation was not merely a separation of the races but *always* an oppression of one and an elevation of the other. As both Bigger and Mary sadly discovered, an impersonal structure could annihilate.

I believe that the slow ascendance of this idea, primarily through the civil rights movement in the fifties and sixties, completely realigned America's social morality. There had been the illusion, especially in the South, that one's prejudices were benign, one's hatreds a private and harmless affair if not acted on. Structuralism gave America a vastly more rigorous social morality by making the point that anything less than fighting against prejudice amounted to complicity with it. In fact, structuralism pushed social reform ahead—it really made the civil rights movement possible—by expanding the concept of moral complicity to include passive attitudes and even silence. These, too, could facilitate oppression.

But structuralism goes too far when it projects a monolithic determinism on the world. Once in the grip of this idea, especially where race is concerned, almost anything can become a structure that automatically "victimizes" certain classes of people whether it intends to or not. For example, a structuralist might argue that white flight is a social structure that oppresses blacks by abandoning them to a low tax base, terrible schools, and violent neighborhoods. The standards of merit used by university admissions departments are commonly seen as structures that "invisibly" do the work of racism by excluding too many black students. The enthusiastic structuralist thinks of himself as sophisticated and "progressive" because he doesn't waste time with individual racists or even individual victims. He believes passionately in two principles: that victimization is totalistic because it is largely structural, and that "disparate impact" (situations in which blacks are shown to be worse off than whites) always proves that racist structures are at work. If a policy—say a university admissions standard—negatively affects blacks more than whites, the disparate impact on blacks can always be explained by the structural victimization of blacks. Thus the policy (the academic standard in this case) becomes an oppressive *structure* where blacks are concerned.

Despite its excesses, however, the idea of structuralism extended moral accountability in relation to racism and gave the United States a greatly expanded sense of its evil. By showing how mere attitudes could be oppressive determinisms, it made the point that no one was innocent. But there is an inevitable companion idea that emerges whenever structuralism is widely accepted. It follows from the simple logic that if bad structures cause oppression, then good structures can bring good things like minority advancement. Thus *interventionism* is the idea that good and wise societies can intervene and fix things like inequality with structures that determine equality.

Interventionism is the activism that follows from the insight of structuralism. It is self-conscious, proactive reform. If inequality is delivered by structure, why shouldn't we erect good things like "diversity in the workplace" or "multicultural atmospheres" or "opportunity structures"? Thus interventionism became associated with the nation's effort to redeem itself from its history of racism. It also became virtually synonymous with black advancement in American life.

In fact, inteventionism's unassailable good intentions have caused it to be seen by many as the very expression of social virtue. And yet, for me, it was precisely this virtuous interventionism that over time began to feel more and more humiliating. It gave me a Bigger Thomas problem. Whether the determinism was bad, as in the case of Bigger, or intended to be good, as with interventionism, blacks were still seen as determined beings without will or agency, and therefore without full humanity.

James Baldwin was contemptuous of Wright's *Native Son* in a famous essay called "Everybody's Protest Novel." He thought Wright had created an inarticulate, nearly subhuman black merely to show the deterministic power of social forces. Of course Wright's good intention was to protest those forces. But for me it was Baldwin who had the more telling last word: "The failure of the protest novel (here I would insert interventionism) lies in its rejection of life, the human being, the denial of his beauty, dread, power, in its insistence that it is his categorization alone which is real and which cannot be transcended."

3.

A month or so before my lunch with the black woman journalist I mentioned above, I had had another lunch in the same restaurant with a white male journalist who was writing a book in favor of affirmative action. Apparently it was a season of lunches and journalists. Like myself this man was an aging baby boomer, so there was some generational kinship between us.

Our actual discussion of affirmative action went nowhere. It was a debate of bromides for and against group preferences. I thought he argued as if from a script, but I probably did too.

"Why do you care?" I finally asked. "Why such passion for an idea like group preferences?"

"Because I think they—" he paused. "A little judicious use of preferences can do great things, can bring a healthy diversity, that's all I'm saying."

"Such passion for a device?"

"Because it's a good thing . . . a good thing. . . . People believe in it. Corporations, the government, universities . . . a lot of people believe . . ."

I will confess that I did not much like this man. But the reason had less to do with him than the fact that group preferences had come to be the conventional idea of social responsibility in America. And because they had become a conventional idea of reform, people did not support them so much as *conform* to them.

My problem with the journalist was my feeling that he was essentially a conventional man. By this I mean that he took a democratic view of the truth—the popularity of an idea was an indication of its truth. He was not supporting group preferences because he had critically considered them; he was *justifying* his conformity to a conventional idea of social virtue, one in which he had much company. And it was the company that counted. He gloated a little as he mentioned that affirmative action was "de rigueur" in the corporate world. And when he uttered the name "Colin Powell," his eyes narrowed almost into a taunt. After all, conventions survive by the number and quality of their endorsements.

Still, his support of this convention had no resonance with what had to have been his inner beliefs. Would he have encouraged his own children to overcome a deficit by looking for a preference? Did he think a preference built esteem or undermined it? Did he believe that in a democracy race should establish different categories of citizenship? When race was at issue, he suspended the usual wrestling with self and deep beliefs that strong convictions require. Mere expedience sufficed. Why?

Because, since the sixties, when the nation was made accountable for its history of racism, white Americans have been without moral authority where racial matters are concerned. The dilemma for the white liberal has been how to display racial virtuousness while lacking the moral authority to assert his or her own truest beliefs and values. Where blacks were concerned the liberal could stand for an engineered racial equality but not for the principles of merit, excellence, hard work, delayed gratification, individual achievement, personal responsibility, and so on—principles without which blacks can never achieve true equality. At home, where his moral authority is sound, the liberal no doubt emphasizes these principles to his children as values in their own right, but where blacks are at issue this same emphasis "blames the victim." So, in this public racial realm, conventional notions of racial virtue are important because they put the liberal in a virtuous crowd, and the crowd compensates for the lack of moral authority. For this reason a look of conventionality, of seeming to represent an indisputable moral norm, is essential to all liberal positions on race.

And what makes structural interventionism so appealing as a conventional idea of reform is its promise to make social virtue automatic, pervasive, and impersonal. The intervention would serve as a kind of automated social activism, invisibly doing the work of social justice—and automatically restoring American moral authority. To have more college-educated minorities we don't need to work at instilling the principle of intellectual excellence, or at raising the standards in inner-city schools, or at making minority neighborhoods safe for children. (In fact, we allow license and lowered standards to prevail in these areas.) And we don't need to engage our "client population" personally. A group preference in college admissions is a simple and impersonal intervention by which we can manufacture a wonderfully "diverse campus"—even when black students average three hundred SAT points below whites and Asians, as has been the case at the University of California at Berkeley.

The generation the journalist and I both come from was supposed to have a higher sense of social morality than earlier ones. I won't argue this one way or the other, but I do believe that we've had more moral vanity than either our forebears or our progeny. The many postwar blessings we enjoyed burdened us with high expectations of specialness. And, having no more innate specialness or talent than other generations, to some extent we turned social virtuousness into a kind of talent, into a legitimate form of specialness with which we might meet our generational mandate to be special. After the Civil Rights Bill of 1964, it was easy for us to place ourselves on the right side of race and gender inequities that had festered for centuries, to borrow a little gravitas, as it were, from the nice timing of history, which gave us these conflicts when they were already essentially resolved.

Over time, structural interventionism, with its promise to automate social virtue, and its flight from the difficulty of timeless principles, became irresistible to our generation. After all, we were working within a society that had just lost its moral authority and, against this disgrace, interventionism looked like hope itself. Principles themselves had been disgraced because so much racism had been carried out by using them as a rationale. Thus relativism became a virtue as a counterpoint to the disgrace of hard principles, and as a way of justifying "innovative" interventions like group preferences that blatantly stepped on principles.

And so, here was this boomer generation in the seventies, eighties, and nineties relentlessly moving through one American institution after another, unfettered (in racial matters) by demanding principles, made special by historical blessings, given a free hand in social reform by the collapse of moral authority all around us. And, as if this were not enough, we were also given an ideology of relativism by which we could weave a glib coherence through all we did. Structural interventionism—as an abstraction, an ideology, a science—was a *reflection* of the almost inevitable vanity of this generation. Its angle of approach on social problems—downward from insight to application, from theory to manipulation, from proposal to program—would define our angle of approach to intractable human problems. In other words, structural interventionism defined our generational concept of social responsibility. It became a theme of the generation's identity. And our love of government was really an identification with its authority to introduce structures, and therefore to implement the good society.

I don't believe that this journalist, in the book he was writing, was alarmed only about the recent attacks on affirmative action. These attacks resonated as a broader attack on structural interventionism, and therefore on social virtuousness as he knew it. They also accused him on a personal level, not so much of moral corruption as of moral glibness. While in full pursuit of his career in journalism, concentrating on all its demands and rewards, he had supported a concept of social reform that would assuage his social responsibility without sacrifice on his part. He could win true virtuousness merely by holding a view, by being in favor of a policy. Affirmative action had not engaged him; it had spared him. It was reform that demanded no principles from its beneficiaries and no sacrifice from its supporters—the fruit of a liberalism with little moral authority but much moral vanity.

To uncover affirmative action, to look carefully into the suppositions on which it rests, is also to unveil at least one theme of my generation's vanity—our general willingness to have the glib, "innovative" idea stand in for principle and difficult struggle.

4.

American social scientists are not immune to the same lack of moral authority in the area of race that affects the rest of the country, and this deficit has severely undermined the objectivity of research on racial matters. Especially after the storm of outrage over the famous *Moynihan Report* in the mid-sixties (the first work to call attention to patterns now associated with the black underclass), many white social scientists backed away from race-related research altogether. Those who went forward (white or black) tended to be liberal, and their work reinforced the blacks-as-victims convention through a very specific pattern of research that I will call specimenization.

This kind of research does essentially four things: First, it makes structural interventions by the society the contingency on which the resolution of some black problem depends. Second, it always establishes the helplessness of black people in the face of the problem being studied. And this helplessness sets up the third characteristic of specimenization research: the assertion of a relationship of obligation in which the larger society is assumed to be responsible for solving the black difficulty. Thus the well-being of blacks is contingent on interventions for which the larger society is responsible, because blacks are—for whatever reasons—helpless to solve the problems they face. Black helplessness is the "given" that justifies all the other suppositions of this research. Therefore the fourth characteristic of specimenization research is that it always builds a case for black helplessness at the expense of black humanity. It shows blacks as "in over their heads," as lacking the internal resources to be responsible, as demoralized to the point of finding hope only by putting their fate into the hands of other people.

The specimenizing social scientist is not a scientist so much as a methodological advocate—someone with an a priori commitment to structural interventions that he will support not by objective study but by studying only those variables of a problem that support intervention. He assigns contingency, obligation, and helplessness before he sets to work. Thus he is really an ideologue working under the imprimatur of social science.

There is probably no more classic example of specimenization research than the work of William Julius Wilson, particularly his last books, *The Truly Disadvantaged* and *When Work Disappears*. When Wilson claims in these works that the loss of industrial jobs in the inner city is the overriding cause of the collapse of inner-city black families, the growth of the black underclass, the proliferation of social pathologies, and so on, he is not giving us the results of objective research. He is offering an ideological and predetermined arrangement of causality, one that is designed to support a demand for interventions from the larger society. Wilson collects and sifts through great masses of data on poverty and inner-city life, much of it revealing, but he establishes no direct causal link between joblessness and the many problems he ascribes to it. He uses no control groups. He does not isolate joblessness as a variable from other variables (welfare without a time limit or work requirement, for example). Instead of rigorous science he works by rough inference and unexamined correlations to support joblessness as the all-determining variable he wants it to be.

Of course, any variable a social scientist chooses to study in the area of race will have clear ideological implications. The controversial inference of Richard J. Herrnstein and Charles Murray's *The Bell Curve* is that the IQ gap between blacks and whites might be roughly 60 percent genetic. This has far-right and, I have to say, even fascistic ideological implications. If genetics contribute this much—or at all—to the IQ gap between whites and blacks, then racial inequality is more a reflection of nature than of white racism. The state cannot remediate a genetically determined inequality, and so by implication is absolved from precisely the interventionism that Wilson's work advocates. Like Wilson, Murray and Hernstien study the variable that suits their ideological purpose, which is to undermine big-government liberalism.

In racial matters social scientists wage a kind of war of variables that is, in fact, ideological warfare. And it is the pressure of this kind of warfare that leads them very often to play a little loosely with the distinction between inference and hard causality. Particularly in the study of race, the kind of hard and absolute causality that we routinely require in the hard sciences is simply not possible. When we don't even have agreed-upon definitions of things like race and intelligence, for example, it is impossible ever to purely isolate them as variables. The best the social sciences can hope for in the area of race is a very gross isolation of variables that *infers.* But inference poorly serves ideological struggle, in which people long for hard causalities to throw at the other side. Also, because the nation is committed to its racial redemption, there is great pressure for the social scientist to discover interventions that uplift blacks. Careers, grant money, and prestigious positions hang in the balance.

All this brings out a certain entrepreneurialism in many social scientists who work in race-related research. One has to build a constituency for one's variable, raise grant money to support it, fashion it into a "promising" intervention, propose public policy around it, and so on. And when it is difficult to isolate this variable with genuine integrity, smoke and mirrors can be used to make an inference seem stronger than it is. An effect achieved by an unmentioned variable can be attributed to the variable the social scientist is arguing for. (It could be reasonably argued, for example, that welfare as a variable generates precisely the effects that Wilson attributes to joblessness.) Control groups that favorably highlight the variable one is pushing can be used whether or not they are true control groups. (Harvard University's John Witte, who was charged with researching the effectiveness of the school voucher program in the Milwaukee schools, has been accused [*The Public Interest*, Fall 1996] of supporting the enemies of vouchers by manipulating control groups to show that voucher users have not gained academically.)

However, the most important entrepreneurial device that liberal social scientists use—and most race-related research is conducted by liberals—is specimenization itself, which contextualizes their variable so that it has a social and moral importance that the science behind it would never warrant. By showing blacks as helpless, their fate contingent on white intervention, and society morally obligated to them, the liberal social scientist can make his or her chosen variable seem literally redemptive. Specimenization injects America's tragic racial history into the science, so that the science comes to have a moral as well as scientific standard, and so that we evaluate the variable being studied for its redemptive potential as well as for its scientific validity. The specimenizing social scientist tries to seduce society's goodwill by presenting the variable as an opportunity for good works. And this is the sort of excitement such research generates, not the excitement of scientific illumination but that high and hopeful mood that precedes social activism.

In the first half of this century about 50 percent of the entire black population of the United States left the poverty and racial enmity of the South for the possibility of greater opportunities in the North. Known as the Great Migration, this was the kind of adaptation that human beings have always made. Why do so many blacks today remain in northern inner cities decades after the evisceration of the job market? In some cases, as Wilson points out, the jobs have moved no farther away than the suburbs, but these descendants of the Great Migration have not followed them there. Why not? Why do uneducated immigrants with little or no English come into these same jobless and blighted neighborhoods, where Wilson and other liberal social scientists say there are no "opportunity structures," and thrive? If education is so obviously needed to join "the new global economy," why isn't education pursued as the northern "promised land" once was? Why are local schools not cherished as portals of opportunity? And, given the high crime rates, poor schools, rampant violence, drug trafficking, and general mood of hopelessness, why don't people leave as they once left the South? As bad as these factors are, they are no worse than Klan violence, disenfranchisement, hunger, disease, and the daily humiliations of racial segregation. Why does suffering not motivate this population the way it motivates others?

In a recent issue of the *New Republic* the black cofounder of the rap group Public Enemy, Bill Stephney, says, "It is not just the demise of work in urban America that has alienated black men from the family-supporting and child-rearing positions they used to occupy with pride, it's a welfare/child-rearing system that has substituted for them." Stephney goes on to talk about why young inner-city women don't get married.

> Marriage will knock out the child-industrial complex she can enjoy, which can include AFDC, health benefits for her children, housing and energy assistance programs, day care, off-the-books employment such as babysitting, hair weaving and back-up singing, cash from new and old boyfriends, cash from mother and relatives, cash and gifts from her homegirls. . . . This "CREAM" scenario—"Cash Rules Everything Around Me," in the words of rappers Method Man and the Wu-Tang Clan—is all these young men and women know. We have the money-for-nothing welfare mentality to thank for it.

Isn't it at least possible, if not likely, that welfare has diffused the normal human impulse to migrate from joblessness toward opportunity, the same impulse that once drove the Great Migration? When, as Wilson puts it, "work disappears," the common human response is migration (or some other radical adaptation), not the inertia of today's inner cities. Inertia is the common response to situations like welfare, where just enough basic needs are met to undercut motivation and risk taking. Why does Wilson reverse human nature itself by making joblessness rather than welfare the preeminent cause of inner-city inertia? Conversely, why can someone like Bill Stephney, a rapper and businessman, see so vividly what one of America's preeminent social scientists overlooks? I think the obvious answer is that Stephney has lost patience with the *convention* of specimenization, which relentlessly shows blacks as helpless. He sees blacks not as helpless but as full human beings who are corrupted and debilitated by welfare, their ingenuity going into petty schemes. He blames society for engendering this "money-for-nothing welfare mentality," and blacks for adapting to it rather than seeking and *creating* larger opportunities.

Specimenization disallows precisely this angle of blame. It has society as the only permissible object of blame. So Wilson, who scrupulously studied the inner city for decades and at the expense of millions of dollars, cannot finally offer the frank insight of a businessman with no training in the social sciences. No matter what his research finally uncovers, the convention of specimenization determines its shape and meaning. For the sake of credibility the liberal social scientist may mention black responsibility but will never seriously push it as the first means to black uplift because *interventions* are his or her predetermined goal. In the end such work will always show blacks as helpless, their fate contingent on white intervention, their weaknesses the result of social determinisms. This is the ideological map of liberal social science in the area of race, and Wilson's subscription to it makes him, paradoxically, more the ideologue and less the scientist than the frank Stephney. Black helplessness is the sacred ideological creed of race-related social science. Even to qualify it in a serious way is to blame the victim.

5.

At a friend's house for dinner not long ago, I met a newly retired doctor who shook my hand and forthrightly announced, "I am very proud to be a liberal." Though his smile defused some of the challenge in this statement, it did not defuse it all. Someone had told him I was a conservative, and this seemed genuinely to disturb him. He had a great shock of pure white hair, wire-rimmed glasses, and, except for his fierce blue eyes, looked like Hollywood's idea of a friendly country doctor. At dinner he sat next to me and, as the salad arrived, he began to make his case for liberalism. But it was not a theoretical case, not an argument for a utopia of some kind. Right from the start it was essentially a case for black helplessness.

He argued for this as if for his own decency. And it was axiomatic to his argument that blacks were unable to achieve true equality with other races on their own. Their helplessness was the result of, and proof of, a great human indifference in the American soul. And this was the second theme of his argument, this idea of the United States as an instinctively oppressive country. There was this blight of racism and black suffering, and then there was the government, the only "reliable force for compassion" that we had. At least three times he told me that Colin Powell would never have made it without government-sponsored affirmative action.

Even Colin Powell's success was, for him, *contingent* on the intervention of a compassionate agent. And I would not have minded this so much had there been at least some recognition of Colin Powell's talents and abilities. But he spoke as though Powell had merely been levitated to great heights by interventions that good Americans like himself had insisted on. In his formulation of liberalism, black people were inert and invisible. And they faced a racism that was so monolithic and impenetrable that they could not be active agents in their own uplift.

His liberalism did not come from a human identification with black people. For him there was the racism in America's soul and then the interventions to contain it. The unintended consequences of these interventions for the people they were designed to help were not a consideration. After all, he made it clear that he had never "personally known" many blacks in the first place, so he would not have been sensitive to whatever unintended consequences they endured. I think it was primarily America's racial shame that troubled him, and that he felt diminished by. Interventionism was his passion because it was an action against this shame that joined morality and ingenuity in a way that made a "structural" moral activism possible. But, from where I sat, it also left him with a familiar white liberal dilemma: What redeemed him by positioning him against the evil of racism also had many debilitating effects on the people who had suffered from racism in the first place.

A fundamental weakness of post-sixties liberalism has been its greater preoccupation with national redemption than with what it actually takes for blacks to achieve self-sufficiency and equality. The great ingenuity of interventions like affirmative action has not been that they give Americans a way to identify with the struggle of blacks, but that they give them a way to identify with racial virtuousness *quite apart* from blacks.

None of this is to say that the doctor sitting next to me at dinner that night was a hypocrite, or that white liberals generally are hypocrites. The doctor was old enough to have lived half his life in the segregationist era. He would have had knowledge of himself benefiting from segregation or looking away from it or rationalizing it. And added to this would be that peculiar late-twentieth-century knowledge of the extraordinary human devastation that simple complicity can make possible. It is a mistake to think that only blacks truly know segregation. Whites know it too, not as its victims or even very many as its open perpetrators. But it made whites know—on some level—how simple a thing evil is, how ordinary, how close at hand, how compellingly convenient it can be. This kind of knowledge makes for its own urgency—an urgency coming out of the white experience of segregation and racism. And in the United States this is the urgency that parallels, but then diverges from, the urgency that blacks feel for full equality. Two experiences of American racism, and two kinds of redemption needed—one from the shame of living with or practicing racism, the other from the shame of being subjugated by it.

I believe it is the former urgency that drives the liberalism of people like the doctor. When he announces provocatively that he is a liberal, when he defends interventions like affirmative action on mere faith, he argues that it is to help blacks, but of course it is really the other redemption he wants, the white redemption. His sense of urgency, and his impatience with me, come from a white pain and knowledge. And when he mentions in passing that his mother was a racist, and that the very neighborhood we are dining in had restrictive covenants against blacks and Jews when he joined a local medical practice, I understand that he is telling me, without saying so directly, that I don't know how close evil is. He feels an accountability to that evil. And he seems almost to be saying that interventions like group preferences are not just for blacks and don't have to work just for blacks.

A great confusion in American liberalism after the sixties comes from the fact that the white mandate for redemption can only fulfill itself through a concern for black equality. This has given us a liberalism that treats black equality more as a means to white moral authority than as an end in its own right.

So we often end up evaluating racial reform more by its usefulness to the moral profile of whites than by how well it develops blacks. Universities across the country offer admission preferences to black students, yet this student group has the highest dropout rate and lowest grade-point average of any student group. If black equality were truly the goal, wouldn't policy focus on educational development before college? And if equality of performance between blacks and whites were the goal, wouldn't blacks be held to the same standards as whites precisely so they could achieve that equality?

This same white need for moral authority has also given government and other institutions an obsession with an equality of results. An equality of colors and numbers, a language of "diversity" and "multiculturalism," lets institutions engineer the beautiful picture of equality while pushing aside the black need to develop it.

But then, once in the color-and-numbers game, the full and complex humanity of blacks—who they really are and what they really need—becomes inconvenient. And this is where the pursuit of moral authority ends in something both pernicious and paradoxical. In the world of interventionism, with all its schemes, formulas, and structural manipulations, blacks are relegated to that most alienated of human categories, "the other." Here they are seen as a different kind of humanity, as essentially unlike "mainstream" white humanity. And the essence of this "otherness" is their injuredness and helplessness. Because the interventions are justified by, and respond to, only these qualities, helplessness becomes the identity they are recognized for. It is the identity that makes them useful in the larger drama of white institutional redemption. In a sense otherness and inferiority now bring entrée where they once caused exclusion. But in both cases—white racism or white redemption—blacks remain largely invisible beyond a presumed inferiority.

In post-sixties liberalism it is virtuous to be tolerant of black weakness and to think of blacks as "helpless others" as a way of acknowledging the historic evil of white racism. In other words, this liberalism tolerates black weakness and inferiority because they are the *result* of white evil. The liberal who has high expectations for his or her own children often feels that he or she cannot "push the issue" with blacks. The white mandate for redemption pressures the liberal to tolerate what holds blacks down. And, in this circuitous way, this liberalism endorses a kind of racism.

Double standards, preferential treatment, provisions for "cultural difference," and various kinds of entitlement all constitute a pattern of exceptionalism that keeps blacks (and other minorities) down by tolerating weakness at every juncture where strength is expected of others.

A question I did not think to ask the doctor that night was whether he would have announced himself so provocatively as a liberal if he'd been warned that he was going to meet a white conservative. Somehow I doubt that he would have revealed his politics, or gone on about the power of racism and the helplessness of blacks if I were a white conservative. It was the idea of a *black* conservative that provoked him. And well it should have. The idea of a black openly outside the framework of liberalism is still odd in the United States. Such a person seems to be disqualifying himself from the fruits of America's struggle for racial redemption, standing against his own racial self-interest. And so the doctor argued *for* black helplessness and the ubiquity of racism as a way of informing me of my self-interest—and perhaps of protecting his own. He said, in effect, that in a context of black helplessness and white racism, preferential treatment is a form of fairness. It offsets the inferiority of one race and the evil of the other. A practical sort of justice.

And he was right, as long as the self-interest of blacks is defined by the self-interest of institutions that want redemption. I believe that preferential treatment is essentially a white liberal idea of black self-interest that serves institutions by letting them practice exceptionalism with blacks. The institution gets its virtue-credit, but blacks have their weakness tolerated rather than their strength rewarded. Then, after black weakness has been massaged, accepted, understood, and felt for, people wonder why the infamous gap between blacks and whites on tests and other performance measures won't close. The answer, of course, is that nobody seriously *asks* that it be closed. A defining paradox of post-sixties liberalism is the symbiotic bond between the moral authority of America's institutions and black inferiority.

It didn't help matters for me that the doctor could point to the entire civil rights leadership as supporting interventionism— his practical sort of justice. From the abolitionist era to the present, the terms of racial reform in America have always been set by a coalition of white liberals and black leaders. And since the sixties, interventionism that would engineer blacks to equality has been the virtuous idea of this coalition. But, in supporting interventionism, I think the black leadership has forsaken the black mandate to achieve true and full equality with all others for the perquisites of interventionism—the preferential patronage of jobs, careers, grant money, setasides, diversity consulting businesses, black political districts, and so on. The black leadership, which could have emphasized antidiscrimination and black development as the road to black advancement, chose instead to rely primarily on group preferences and entitlements. This bargain has transformed the civil rights establishment into something of a grievance elite, largely concerned with turning the exceptionalism practiced by institutions in regard to blacks into the patronage of racial preferences.

Of course it is true that interventionism is what white society offered blacks (rather than serious antidiscrimination and development) because the redeeming look of equality could more easily be engineered this way. And the black leadership, coming out of segregation, in which whites had never given much of anything to the black cause, quickly grabbed up interventionism as a valid way to equality. Thus, for entirely opportune reasons, this post-sixties coalition of white liberals and black leaders made equality into a near perfect expediency with no relationship to a human equivalency between the races. As such it could be manufactured without the actual development of blacks to equality.

This meant that the doctor spoke with the authority of the civil rights leadership on his side. It also meant that he spoke with more official "black" authority than I did. And this authority confirmed for him that interventions were the only road to white redemption. Worse, there was the implication that if he sided with me—if he subscribed to antidiscrimination and black development over interventionism—his redemption would be withheld. And, by the odd mathematics of American racial politics, he might thereby be counted a racist. This, of course, is the white liberal's crucible—he gets to define America's racial reform as interventionism, but he lives without even enough moral authority to declare himself racially innocent and have the declaration stand.

So when a white liberal and a black conservative meet, there isn't much business to be done. And the problem is not just in our different mandates. For example, I not only admire the white mandate, but I also admire the white liberal for recognizing it and taking it seriously. This is what I admired in the doctor, his acceptance of this mandate, his understanding that history had given white Americans the responsibility of overcoming racism. What I didn't admire in him—and post-sixties white liberals generally—was the moral self-preoccupation. This is what made them dangerous to blacks—ready to give them over to an "otherness" in which nothing is expected of them. The liberal may feel that the black conservative doesn't give him credit for his moral sensibility, but this is not true. The black conservative appreciates the sensibility but resents the moral selfishness.

One of the great attractions of "conservatism" for blacks today is the freedom it offers from yet another white mandate—not white supremacy but white redemption.

History imposes these mandates on whites and blacks in the form of responsibilities that individuals in each group carry as a part of their racial identity. They are, I believe, absolute. They are more often denied than not by both whites and blacks, but even the denial validates their presence. Today the mere knowledge of what whites did to blacks in history makes whites responsible for showing a moral superiority to their race's behavior in the past. The doctor I met, for example, must show himself far beyond the racism of his mother. Correspondingly, the mere knowledge of an historically imposed inferiority makes blacks today responsible for showing an equality of achievement that their forebears were prevented from showing. History defines our identities as much by the collective responsibilities it imposes as by the selected tales of glory we take pride in. And these two mandated responsibilities—white redemption and black equality—may be unwelcome and will often seem irrational, yet they are at the heart of what it means to be white and black in the United States.

But all this is a problem for the white liberal because he distinguishes himself from other whites by the *intensity* of his responsibility to the mandate. This intensity is his identity as a white liberal. He is not just accepting of this responsibility, he is passionate about it. Passionate moral responsibility is his trademark. Probably this is because passion seems the only correct response to great historical injustice, the least one can offer in payment for redemption. But the problem with an identity based on passion is that it often puts one at odds with reason and common sense.

However, an important qualification must be made here. Though the liberal identity calls for passion, real social passion is all but impossible to sustain over time. And so, like religious fervor, it must be codified into manners and practices that the liberal can "genuflect." These genuflections, then, are the ritualized display of passionate responsibility to the white mandate of redemption. And this, as it were, is not a bad definition of "political correctness." The great problem this poses for liberalism, as for religion, is that when the original passion is reduced to genuflection, it is achieved more by mere conformity than by difficult effort. This introduces the same hollowness into liberalism that is the bane of organized religion—passion as conformity, iconography, and empty observance.

This has made post-sixties liberalism essentially a *received* doctrine, more autocratic than democratic. Amorphous and empty ideas like multiculturalism and diversity do not exist to be defined or debated so much as affirmed as received expressions of virtue. When the California regents voted to end group preferences at the University of California, the president of the University of Michigan announced that he would resign if preferences were voted out on his campus. I think this was more genuflection than reflection, more obeisance than deliberation—the testimony of a man operating inside a received liberalism in which all is resolved and only affirmation is left to him. With no fear of having to back up his threat, he was able easily to display his passionate responsibility to the white redemptive mandate.

The ritualization of liberal passion has hardened it into a brittle mask. CEOs, foundation presidents, government officials, educators, politicians, union leaders, and the man and woman in the street—all can wear the mask of the racial progressive. In a received liberalism of genuflection true reform itself is simply not necessary.

One of the serious problems that this overall rigidity has brought to post-sixties liberalism is a confusion between its broad mandate and the means it uses to achieve it—that is, the assumption that a given intervention is the same thing as the white mandate for redemption. And from this confusion comes the liberal tendency to fight for interventions as though fighting for redemption itself, never realizing that there might be many routes to the fulfillment of this mandate.

I will call this confusion the "Glazer trap" in honor of the social critic Nathan Glazer, a man whose work I have often admired. Back in the mid-seventies, in a famous book called *Affirmative Discrimination*, Nathan Glazer launched one of the first thoughtful attacks on affirmative action. He was ahead of his time in arguing openly that affirmative action had moved dangerously from "equal opportunity to statistical parity." Like other neoconservatives, Glazer was a liberal who had become disenchanted with affirmative action as an intervention. He was not against the white mandate per se, the idea that restored moral authority in the United States would require serious racial reform. His objection, quite reasonably, was to the intervention of affirmative action as a *means* to that mandate.

But, flashing forward twenty years to the nineties, we see that the world has begun to catch up to *Affirmative Discrimination*.

Now preferential affirmative action is under serious assault. The state of California has voted against group preferences in the nation's first referendum on the issue. Exit polls from that election day show that almost 30 percent of blacks voted against preferences. And suddenly, in this atmosphere, Nathan Glazer came out in favor of preferential affirmative action, the self-same man who had launched the movement against it. Why? The reason he gives in an odd *Wall Street Journal* op-ed piece is that the withdrawal of preferential affirmative action would constitute a "rejection" of blacks. Elsewhere he says that the number of black students would decline significantly at elite universities. If this is not racism, it is certainly paternalism, since the first reason presumes to protect black "feelings" and the second accepts an apparently permanent black inferiority that will always have to be accounted for.

However, I sense another reason behind Glazer's recant: that he has confused the specific intervention of preferential affirmative action with the white mandate to win redemption through racial reform. Glazer forgot the very distinction between mandate and means on which his own book was predicated. What made that book important was the fact that it was not written by a person who rejected the white mandate. To the contrary, its authority came from the fact that Glazer *was* an old-line liberal who clearly wanted America to restore its moral authority through racial reform, and who wanted blacks to achieve full equality. *Affirmative Discrimination* was a critique of the *means* to the redemptive mandate, not the mandate itself.

The "Glazer trap," this confusion of intervention with mandate, comes from the calcification of liberal passion into mere propriety and iconography. And in the hardened and reflexive manners of contemporary liberalism, affirmative action is not simply one idea of reform among many; it is an icon of white American racial redemption. Moreover, as an icon, people are called upon to genuflect to it, not to examine it. We don't coldly analyze it as we do economic policy because its only real accountability is to white American redemption, not economic prosperity. And that redemption is won through conformity and genuflection, not the effectiveness of the policy. (President Clinton gets credit for supporting affirmative action, not for making it work.) So Glazer, who once viewed preferential affirmative action as only a social policy, today genuflects to it as an iconic representation of white redemption. And in this he looks around, as all conformists must, for justifications of a policy he saw through twenty years ago.

But the "Glazer trap" goes well beyond liberals. Conservatives, too, have let the intervention of affirmative action be confused with the white mandate for redemption. In Congress the Dole-Canady Bill, which would eliminate group preferences in the federal government, was withdrawn by Republicans who had spoken out against group preferences for years. In the 1996 election the Republicans stayed away from the California Civil Rights Initiative (which ended group preferences in the state government) until the very last minute, when it was clear that the initiative was doing far better than their own presidential candidate. Even these natural enemies of group preferences were afraid to take on the icon of affirmative action. And their fear was very precise: If they attacked the intervention, it would seem as though they were rejecting the larger mandate that called for white America to seek redemption through racial reform. Thus they had no way of attacking affirmative action without seeming racist.

The "Glazer trap" is in many ways the larger white American trap. White America has simply lacked the moral authority to confront the liberal fusion of intervention and mandate, of affirmative action and white redemption. Republicans have been no more able than Democrats to say openly that affirmative action is *only* an intervention, a method, and not representative of America's larger will to redeem itself racially.

The reason America needs racial redemption in the first place is also the reason for our brittle post-sixties liberalism and the "trap" it produces. America's racial history has injured the very legitimacy of institutional America where race is concerned. Unable even to change the *method* of reform without seeming to forsake the redemptive mandate means that the United States cannot carry out racial reform with integrity. Two things have not been possible: reform that works inside the framework of democratic principles, and reform that makes the difficult demands on former victims that are actually necessary for them to achieve parity. Post-sixties liberalism has been undermined, and to a large degree corrupted, by having to live within these impossible parameters. Unable to enforce either principles or difficulty, it has had to let the black grievance elite call the tune. And this elite has been quite happy not only to entrench preferential treatment for blacks as the predominant mode of reform but also to use preferences as a kind of patronage to insure its own power.

The lack of moral authority that racial reform hopes to overcome also makes integrity (and therefore effectiveness) impossible in such reform. This leads to the other side of the "trap," in which the liberal's good intentions undermine the good, in which he or she makes a virtue of discarding principles and suspending difficulty. On this side of the trap the liberal pursues racial redemption by embarking on a kind of corruption. Listen to the leaders of America's institutions speak on race—the president of UC Berkeley saying he wants to create "multicultural atmospheres"—and you hear beneath the good multicultural intention the corruption of human engineering, of a university picking and choosing among human beings by skin color alone. A racially sensitive atmosphere created by human insensitivity. Corruption in the service of "good." Without the moral authority to reform by principle, we end up with an insecure, defensive, and often corrupt liberalism, not of reform but of apology.

6.

The liberalism of the early civil rights era, however, was solidly grounded in democratic principles, and was far more preoccupied with freedom than with virtue. It achieved the virtue of ending racial segregation by enforcing freedom, by applying democracy past even the atavistic barrier of race. The government's activism was to send federal troops to the South to enforce democracy against the racial double standard of segregation that guaranteed white supremacy. The mission of this "freedom-focused" liberalism was to give individuals the freedom to agent their own lives regardless of their race, class, or sex. From the abolitionists to the suffragettes to the civil rights workers, the struggle of freedom-focused liberalism was *not* to free oppressed groups so much as to free the individuals within those groups—to prevent society from using the group to oppress the individual. This is the central commitment of democracy, this idea that freedom does not truly exist until it is grounded in the individual, and that freedom and individual agency are virtually synonymous.

But post-sixties liberalism is much more concerned with virtue than with freedom because it is driven by the mandate to redeem the nation from its sinful racist history. And in this liberalism the government (and American institutions generally) is the agent of redemption, an initiator of social virtue, not indirectly by ensuring freedom but directly as an agent that conceives of the good and then executes those conceptions. So this unspoken, white mandate to redeem America has essentially added a quasi-religious function to government, has put government into the business of spiritual transformation. Moreover, it has made for a highly ideological liberalism that asks all American institutions—corporations, foundations, retail businesses, universities, and so on—to add a redemptive responsibility to their usual functions. Now virtually all of institutional America is accountable for the nation's racial redemption and must help to agent that redemption.

In this sense, what I have called post-sixties liberalism is better termed *redemptive liberalism*—not a discipline of individual freedom but an ideology of conspicuous racial and social virtuousness.

In a 1997 column the television correspondent Cokie Roberts and her journalist husband, Steve Roberts, admonished the black conservative congressman, J. C. Watts, in a very interesting way. After praising Watts's Republican response to President Clinton's State of the Union speech for its focus on "black individualism and self-reliance," they quickly reversed field. They said that Watts, like Colin Powell, was "lucky" to have had "hardworking parents," but that, unlike Powell, he had forgotten "that many African-Americans have not been so fortunate." They then went on to mention a few helpful but "modest" government programs like Head Start and enterprise zones. Of course, we were never told whether or not Watts actually objected to these programs. Nor were we told that he has spent much of his time in Congress working with Rep. Jim Talent on a community development program for inner cities (based on the values he espouses) that is far more ambitious than the programs mentioned by the Robertses. And, as is so often the case in redemptive liberalism, the Robertses gave no evidence that any of the programs they like actually work. (Many studies show that the beneficial effects of Head Start disappear by the third or fourth grade, and most successful enterprise zones have been on the perimeters of cities and not in poor inner cities.) What J. C. Watts truly believed, what he was working for in Congress, which programs actually worked and which didn't—all this did not matter to the Robertses.

What apparently did matter above all else was that blacks once again be seen as helpless, their fate contingent on the interventions of government. In other words, the Robertses' redemptive liberalism specimenizes blacks just as liberal social science does. Watts's sin was that he did not specimenize his own people, though this did not mean that he thought government had no role to play. But his mere assertion of "black individualism and self-reliance" threatened to undermine the sacred foundation of redemptive liberalism—this idea of the "helpless other," this wretched category of humanity in which the expectation of self-help only "blames the victim." Unwilling to ask directly for liberalism, the Robertses instead asserted black helplessness, liberalism's most sacrosanct justification. And, rather than make their own case, they brought in the ubiquitous Colin Powell and, in effect, ascribed to him a wise acceptance of black helplessness. "Powell knows that [black helplessness]; Watts is still learning it."

When redemptive liberals argue, moral authority is the only logic that truly matters. So the Robertses reduce blacks to iconic representations of a point of view—Powell is black helplessness, Watts is black self-reliance—and then identify themselves with the icon that has the most moral authority.

Because the redemptive liberal is cowed by his own lack of moral authority as a white, he is far hungrier for it than for either logic or truth. He is constantly on the lookout for the black with the most moral authority who also represents his point of view. When he finds that black, he can then let the *black*'s authority carry the argument. To combat the significant moral authority of J. C. Watts, a man who rose from difficult circumstances to the U.S. Congress, the Robertses had to reach for the most esteemed black man in the United States, and still it was a close call. On television recently a white male commentator made a critique of affirmative action, but his liberal colleague got the last word when he said that Jackie Robinson's widow supported affirmative action. "And who do you think I'll pay more attention to, you or her?"

The redemptive liberal is not without logic. He is without moral authority. And this is felt as an *inadequacy*, one he or she accounts for by arguing vicariously through the symbolic moral authority of others.

Black helplessness has been the raison d'être of redemptive liberalism, the condition it was born to address, and the best justification for its demands on society. As this liberalism evolved out of the Great Society, it soon generated a new grievance group politics in which racial, ethnic, and gender groups could assert helplessness by claiming victimization. But also it defined social virtue for whites as the condescension of seeing these groups as so plagued by helplessness that the successful among them were "lucky" exceptions who only proved the rule of helplessness. If J. C. Watts made it by self-reliance, the redemptive liberal invalidates this means of advancement by casting him as the exception—a patronizing circularity that makes self-reliance irrelevant precisely because it succeeded.

In redemptive liberalism we blacks lost the first chance we ever had in the United States to truly control our own fate. After the end of segregation in the mid-sixties we made an understandable but profound mistake: We put ourselves, our fate, in other people's hands. This was a very easy mistake to make considering that the president of the United States suddenly offered an entire catalog of Great Society programs that, in addition to ensuring freedom, promised also to restore us. The Great Society was the first ambitious expression of redemptive liberalism. I worked in four Great Society programs from the mid-sixties to the early seventies, and I remember well the headiness of their promise, the feeling that they were virtual incarnations of historical justice and American redemption. There was no way to know the price we would pay, no experience of full freedom to fall back on and to be informed by, no way to understand that the agency over our fate that we had just used to win freedom would be given over to the well-intentioned but confused architects of the Great Society. So this loss of control over our fate was virtually inevitable.

It was also ironic. Just after winning our civil rights in the greatest nonviolent revolution in American history (one of the greatest in all history), we had to turn around and impart to ourselves a degree of helplessness in order to justify the programs of redemptive liberalism. Suddenly a people strong enough to win freedom in a society in which they were outnumbered ten to one had to make a case for their own weakness, had to offer up their own helplessness as a vehicle for the redemption of others, had to reimagine themselves and advertise themselves primarily as victims.

Welfare without a time limit or an expectation of work may have shown white America as compassionate, but it also took the problem of poverty away from those who suffer it. When universities took responsibility for the problem of black underrepresentation on campus, and lowered standards to raise the numbers, then blacks became *invested* in the academic weakness that won the specialness of a double standard. Academic success jeopardizes both this specialness and the favorable double standard. In redemptive liberalism others are responsible for the problems blacks suffer, and blacks are, in an odd way, responsible for preserving the weaknesses that keep others responsible. The black reward follows the display of difficulty, not the display of success.

Redemptive liberalism asks blacks to negotiate so much of our fate through the larger society's hands that our great civil rights victory of the sixties resulted in a curious incongruity: freedom without self-determination.

Redemptive liberalism has encouraged in blacks what might be called a psychology of contingency. I encountered an example of this on my own campus a few years back, when I gave a talk on racial matters to a large audience of students and faculty. In the question-and-answer period we fell into a hot, if predictable, debate on affirmative action, and the auditorium filled with tension. At one point a black professor from the Black Studies Department—a woman I had known for years—rose to speak. Anger had stolen her self-possession, her ability to censor herself, and so out of a kind of general alarm she said: "And if black students do well, they'll end up like the Asians. They'll lose their preference. . . ." Did she realize what she'd said? I asked.

Waiting for the answer, I remembered the countless times she and I had stopped on campus to commiserate as colleagues over some piece of news or some small frustration. And I knew her well enough virtually to see where her wires had gotten crossed. She thought of herself as a kind of mother protector for many black students, and it was this protectiveness that led her to confuse one worry for her students with another. Fearing that they might lose their preference, before she could think, she leaped to caution them against doing as well as the Asians. Such is the irrationality of affirmative-action fights that her supporters only looked fiercely over her shoulder at me as though she had scored a brilliant point. From a group of white liberals across the auditorium (the races here allied in ideology but still divided by color), someone said, "You know what she means."

But it was some time before I would, in fact, see the *meaning* of what she had said. Eventually it became clear that, within the context of redemptive liberalism, what she'd said was logic itself. She saw the fate of black students as more contingent on the interventions of the university than on their own efforts. When she thought of helping her students, when the mother in her lashed out protectively, she did not say that their future was contingent on what they might do or even on how she might help them. It was contingent on the intervention of group preferences. And she would protect those interventions and that contingency, even if it meant encouraging them to avoid the mistake of the Asian students who had lost their preference by reaching for excellence. A thoughtless remark, maybe, but also reflective of the psychology bred by redemptive liberalism, in which the black self is a contingent self that paradoxically preserves enough weakness to ignite the obligatedness of white America.

This paradox—arguing against your own capacity to help yourself as a way of helping yourself—has been a theme of black politics over the last thirty years.

This was the paradox that lent pathos to the figure of Jesse Jackson in the 1996 election, racing desperately up and down the state of California fighting against Proposition 209, the ballot initiative to end group preferences in the state government. Such effort and moral capital in defense of an idea (group preferences) so vulnerable that it could be (and was) voted away in a single day. Here was black America's most visible leader racing from one underattended rally to another, fiercely arguing that black progress was contingent on the willingness of whites to keep group preferences. And what deepened the pathos here, of course, was that this arrangement of contingency duplicated precisely that of slavery, where black fate was indeed in white hands.

What group in human history has advanced by allowing its fate to be contingent on a former oppressor's will to redemption? What group has achieved equality by fighting to *keep* its fate contingent in this way? And yet this contingency literally defines redemptive liberalism, which is based on the promise that the contingency will now be good where it was once evil, will now uplift blacks where it once held them down. So the "good fight" is to lock blacks into it even if it perpetuates a kind of "reformed" white supremacy, a white domination that is benevolent rather than malevolent.

This contingency is the cord that binds blacks and whites symbiotically. And there is near constant trading along its length. Thus it opens a zone of opportunism for both races. This is the zone to which blacks take their problems when they want them to pay off; it is also the zone where whites pay off those problems in exchange for moral authority.

If black children in Oakland, California, are doing poorly in school, we don't simply raise our academic expectations of them and work harder at teaching them; we take the problem into this zone of opportunism that contingency opens between the races, and we say they speak a special black language called ebonics. We then ask for federal money to teach this language to their teachers so they can be more "sensitive" to the racial self-esteem of these students.

Normally when the academic failure of black children is taken into this zone, it quickly ignites obligation in government agencies, foundations, universities, school districts, and the like. Money flows, jobs materialize, careers advance, and so on. When ebonics finally made the national news, several fully funded ebonics programs were already up and running in California and elsewhere. Had ebonics not become a laughing-stock, it might have thrived as a moderately profitable idea. Indeed, there will still likely be profit taking in the form of MLA (Modern Language Association) papers on the herme-neutics of ebonics, linguistic investigations into the African roots of the language, grants given to examine it as a teaching device, conferences, and more. Even its rough ride in the media will likely not banish ebonics entirely from the zone of oppor-tunism.

But ebonics is merely a subcategory of two vastly more effec-tive "contingency triggers" that have been put to yeoman duty over the years in setting up this zone of opportunism—racial identity and self-esteem. (Another term for this phe-nomenon, which I will use in a different context, is *indirec-tion*.) Contingency triggers try to establish the root cause of a black difficulty—directly or indirectly—as racism. When they succeed in this, they tie the solution of the black difficulty to white obligation, thus triggering the contingency. Now the zone of opportunism is open, and the bargaining can begin. Racial identity and self-esteem are extremely effective contin-gency triggers because they are so amorphous. We can say that almost any problem black children suffer from is due to injured racial self-esteem (or not having their identity reflected), thereby linking it to racism and making it contingent on white obligation.

Also, because it was white racism that injured black identity and esteem, whites themselves cannot be employed in solving these problems. Only blacks can do this work. Thus identity and self-esteem not only trigger contingency, they also give blacks a monopoly on the jobs, administrative positions, and research grants involved in solving these problems. In other words they obligate and exclude whites at the same time, so that whites can *fund* but not *fix* black problems.

Culture is another all-purpose "contingency trigger" that pushes black difficulties into the zone of opportunism by tying them to exclusion. Thus multiculturalism serves as an ongoing contingency trigger. Having very little to do with the actual business of culture, it is a construct that implies white exclusion of blacks and other minorities every time it is uttered. It is an idea that claims to exist as a defense against white racism, so it automatically obligates whites to provide separate racial and ethnic territories in order to prove their commitment to inclusiveness.

Also, because racism created the need for multiculturalism, simple justice requires that inclusion of blacks and minorities not be hindered by criteria of excellence. Minority representation has greater moral urgency than does fairness by merit. In fact excellence and merit can be the very arms of oppression. Teachers' unions in California have filed suits claiming that minority teachers are discriminated against by teacher competency examinations pitched at a mere tenth-grade level. Universities routinely argue that test scores and grades for minority students and publishing for minority faculty are not truly relevant criteria. Medical schools often argue that they lower standards for minorities so that there will be enough doctors for minority neighborhoods—despite what this implies about the quality of doctors in minority neighborhoods. Wherever black representation is an issue, excellence is cast as an adversary of fairness.

Multiculturalism, like identity and self-esteem, is a contingency trigger because its primary goal is not to illuminate culture but to obligate institutions to open up exclusively black territories and monopolies, and to remove excellence as a barrier in this process. The unique character of black American culture—its many art forms, its religious rituals, its manners and customs—is of very little interest in a multiculturalism that reduces minority cultures to the theme that best triggers white obligation: victimization.

At least since the sixties, *race* has set the terms by which moral authority is pursued in the United States. Thus it has given us our practical, everyday idea of what social virtuousness is. It has also given us the unique American paradigm of redemptive liberalism in which black equality and white redemption are mutually contingent. This, in turn, has left us with two ongoing obsessions: contingency triggers and interventionism. The former triggers the obligation that is paid off in the latter. Our national debate on race tends to be around the validity of contingency triggers and the interventions they generate. For example, is racism the cause of black underrepresentation in an institution? Is affirmative action a reasonable intervention to address it?

Americans almost never discuss race to understand it better or to diffuse it as a barrier between people. Rather we argue over whether racism is the cause of a black problem. If it is, then the entire paradigm of redemptive liberalism from contingency to intervention kicks in. If it is not, society is off the hook—no contingency, no obligation. These are the terms of the debate to which redemptive liberalism has led us.

But today this form of liberalism applies well beyond race, because its paradigm determines how moral authority is won in many areas of American life. We now cast problems as disparate as the environment, gender relations, education, and consumer safety as injustices contingent on America's will to redeem itself. All these problems are thrown into a zone of opportunism and bargained over. But the problem is that society is rewarded more for seeming to redeem itself than for solving the problem.

So we don't simply go to work on environmental problems; we show ourselves to be redeeming America from a history of cruel imperviousness toward the environment. We don't simply enforce reasonable sanctions against the sexual harassment of women; we redeem society from the shame of male boorishness. We don't correct obvious problems so much as try to establish our innocence in relation to them. Today even diseases like AIDS and breast cancer, which affect groups with historical grievances against the United States, can trigger the contingencies that lead to special interventions (like more research money), which in turn allow America to redeem itself from the "indifference" or "prejudice" through which it no doubt "structurally" contributed to the disease.

So, whether with the environment or even certain diseases, America is always paradigmatically fighting the race problem, redeeming itself from a national shame more than problem solving, asserting its innocence in relation to a problem more than overcoming it.

And this paradigm is also invoked as a means to power. After all, the moral authority that redemptive liberalism is after *is* power. Therefore politicians and countless other leaders can wield this paradigm to gain power. They can trigger contingency by implying that oppression is the root cause of a problem, and by casting those who suffer the problem as helpless. This pushes the problem into the zone of opportunism, where the politician or leader can win moral authority (power) by coming up with an intervention to address it.

When President Clinton said in his 1997 inaugural speech that he wanted to ask one million college students to volunteer to teach reading to America's children, he was using the paradigm of redemptive liberalism as a means to power. He established the nation's poor-reading problem as a contingency by implying that both the students and their teachers were helpless against it. Once the principle parties were seen as helpless, the fate of reading in America was clearly contingent on America's will to redeem itself from this national shame. (Always redemption before problem solving in this liberalism.) He then proposed the stunningly awkward intervention of having "a million" college student volunteers invade America's classrooms to teach reading to children. He even admonished teachers not to be territorial and to accept these college students warmly in their classrooms.

But the president forgot something important: The reading problem of America's children was not caused by oppression or exclusion. Students had not been oppressed into poor reading, and teachers had not been oppressed into poor teaching. Nor was there any evidence that either group was helpless. And, without either oppression or helplessness, there was in fact no contingency. There was no reason for the fate of their reading skills to be contingent on America's will to redeem itself through an elaborate intervention.

If anything, it was the students, their families, and their teachers who had failed America, rather than the other way around. Whatever one's feelings about the funding of public education, there is no doubt that America's schools are well enough supported (nine thousand dollars per year per student in inner-city Washington, D.C., for example) so that it is not too much to ask that basic reading be taught. Shouldn't President Clinton have been demanding more for the country's money? Instead he later made it clear that his "volunteer" program would cost an additional two to three *billion* dollars of public money. But aside from money, what is there to suggest that college students will be able to achieve what trained teachers have not? Since America's schools have successfully taught reading for centuries, and in the face of subsequent waves of immigration, why the sudden need for such extravagant help? And how would schools logistically accommodate one million college students? What training would they have? What would be left for teachers to do?

I think the president was treating America's reading problem as if it were racial oppression because he was exploiting the paradigm of redemptive liberalism as a means to power. He asserted this problem as a contingency that he could be the master of. In effect, he made American children and their teachers into blacks. He veiled them in the same helplessness that blacks are made to carry, so their fate could seem to be contingent on his determination to redeem the nation. And once he had them contingent on himself, he could present the great intervention that would show him to be the master of that contingency—the man who knows the way from despair to hope. So positioned, he stood to gain moral authority, and thus power, *in proportion* to both the seriousness of the problem and the grandiosity of his intervention.

But it was race—and the peculiar liberalism by which America has tried to redeem its racial history—that made this opportunism possible.

In the fifties, when *Sputnik* challenged American schools to improve science education, President Eisenhower did not make the problem contingent on America's will to redeem itself morally. He did not ask for a million college student volunteers. He asked the schools to beef up their science programs, and they did. But race had not yet made social problems into opportunities for redemption. Eisenhower just wanted to get the job done, not to become a master of contingency.

7.

Often it takes the Tocquevillean view of the outsider to see the most defining features of American life. Thus, V. S. Naipaul, the Trinidadian writer, sees a resonant theme in the much-taken-for-granted American idea of "the pursuit of happiness," a theme not only of American culture but also of a new "universal civilization"—by which he seems roughly to mean modernity. Of this "pursuit of happiness" he says: "It implies a certain kind of society, a certain kind of awakened spirit. . . . So much is contained in it: the idea of individual responsibility, choice, the life of the intellect, the idea of vocation and perfectibility and achievement. It is an immense human idea."

This "immense human idea" asks people to discover themselves as individuals by responsibly pursuing their happiness. In a free society the self, the individual identity, the singularity of a person, unfolds in this personal pursuit of happiness. But there are also many rigors involved—the development of skills as one's personal currency, the struggle against limitations that makes this possible, the often difficult exercise of choice, the pursuit of aspirations around immovable realities, the will to endure the high risk that goes with our best possibilities. Happiness in this context is a kind of difficulty—one that the civil rights movement struggled to win for black Americans.

It was always the collectivizing mark of race that kept

blacks from a full engagement with this difficulty and that held them back from the freedom in which, as Naipaul put it, "it was necessary to be an individual and responsible." Racial oppression imposes nonindividuality on its victims, tells them they will achieve no self, no singularity, that will ever supersede the mark of their race. This surely is the opposite of happiness, this confinement of the self inside a color. The early civil rights movement—grounded in freedom-focused liberalism—saw the mark of race as anathema to freedom, to the individual, and to the pursuit of happiness. It wanted freedom from racial determinism. Therefore, it was a struggle *for* the black individual and *against* his or her race as a political determinism. This was how the great movement sought to bring blacks into the difficulty of a true and unencumbered pursuit of happiness.

But then, in the mid-sixties when greater freedom came, the nation changed its preoccupation to redemption and to the proactive reform by which it hoped to show itself redeemed. And here, in the idea of systemic and structural interventionism as a means to black uplift and white redemption, is where things began to go wrong for blacks. Here is where agency over our own fate was traded away, so that happiness was not something the individual pursued but something the group waited for. Worse, these race-conscious interventions once again submerged the individual in his or her race, deindividualized him or her, and, ultimately, obsessed the nation with group identities.

Just when the idea of the individual might have taken hold, the idea of interventionism came in its place—and with it specimenization, helplessness, contingency, and overreliance on white moral obligation. This is where the black individual lost out to the nation's need to redeem itself. And this is also where we became essentially a *sociological people* with a sociological identity, a group moving from the dehumanization of oppression to the deindividualization of the remedies for it.

A relentless theme in the essays of Ralph Ellison is that, above all, blacks should be spared from sociologists. "It will take a deeper science . . . ," he says, ". . . to analyze what is happening among the masses of Negroes." For Ellison blacks were not specimens; they were "personalities of extreme complexity . . . , personalities which in a short span of years move from the level of the folk to that of the sophisticate, who combine enough potential forms of Western personality to fill many lives." Of course, Ellison wrote these lines in the fifties and sixties, before the era of interventionis when blacks began to think of themselves and their culture in the flat, sociological terms that justify interventions—as specimen/victims. His point was that racial oppression had not produced automatons but rather had forced "the Negro down into the deeper levels of his consciousness." Therefore it forced an intense and original individuality, an extraordinary talent for self-invention, for masking, for improvisation.

A self-invented and intense though well masked individuality was an adaptation to an absurd predicament. Thus, in an era of intractable segregation and unbending racial determinism, there emerged a Duke Ellington, whose music and persona defined elegance and sophistication for all of America. And today, at the other end of the aesthetic continuum, an entire industry has sprung up around rap music, the invention of poor, inner-city blacks who grow up without families or educations amid an unimaginable array of social pathologies. Whatever one may think about rap music, the helpless black specimen/victims of social science and racial politics would not be able to create such a profitable cultural invention—one that has shaped the tastes, styles, and metaphors of a generation. And what enabled this invention to flower into profitability was the full array of "conservative" values—individual responsibility, initiative, discipline, perseverance—that redemptive liberalism sees as "victim-blaming" when applied to poor inner-city blacks. It cannot be coincidental that in those areas of greatest black achievement—music, literature, entertainment, sports—there have been no interventions whatsoever, no co-optation of agency, no idea that some "opportunity structure" will enable blacks to participate.

Interventions in the name of race suppress the individual because they always impose a collective expectation. Whether it is the latest educational "innovation" for black students or group preferences, the intervention comes to the individual not because of personal uniqueness but because, as the critic Edward Bland says, he or she is "a member of an ostracized group." The group identity earns the intervention, the attention, and concern of the larger world. The individual (say a black public school student, since this group has been sat upon by thirty years of "innovative" interventions) is expected to respond to the intervention as a specimen of the group: Because interventions by race see individuals as specimens of the race, they encourage people to confuse and even substitute group identity for individual identity.

However, today's black grievance elite *wants* black Americans to be a sociological people. It *wants* us to be a "race" people so that even when we think of ourselves as individuals, it is our race, our group identity, that is paramount. Most important, this elite wants us to frame our problems as sociological problems that we suffer essentially because we belong to this race. Their reason is very simple: Our group identity as black Americans is, in itself, a contingency trigger. Because this was the identity that America oppressed for centuries, it is the identity that made our fate contingent in the first place.

So it is this collective black identity, not our individuality, that obligates white America to us. Therefore the black grievance elite exaggerates the importance of this identity because it is what opens the zone of opportunism between the races. Hence race-based interventions—Afro-centrism, "ebonics," racial and ethnic "learning styles," black all-male academies that hope to compensate for the lack of fathers, group preferences of all kinds, the ubiquitous idea of racial role models, multicultural curricula, and the rest—all spawn hundreds of thousands of jobs, boost countless careers, and bring in hundreds of millions of dollars in public and private money. Race-based interventions support a vast petite bourgeoisie of attendants to the self-image and self-esteem of helpless black "others."

But the price paid for all these interventions is to suppress black individuals with the mark of race just as certainly as segregation did, by relentlessly telling them that their racial identity is the most important thing about them, that it opens them to an opportunism in society that is not available to them as individuals. Black politics, since the sixties, has been based on this hidden incentive to repress individuality so as to highlight the profitable collective identity. The greatest threat to the grievance elite is a society in which the individuality of blacks supersedes their racial identity in importance. The iron law of this racialist elite is that race is contingency and individualism is nothing. So much for the idea of blacks moving into a free and modern society in which "it is necessary to be an individual and responsible."

In universities, schools, corporations, government agencies—environments where there are many race-based interventions like preferential affirmative action, diversity goals, multiculturalism, and so on—there will also be an overemphasis on group unity and conformity, and an intolerance for any individuality that even *appears* to be at odds with the group. However, in areas where there are no race-based interventions, individuality becomes the vehicle to success, and group conformity becomes secondary.

The black athlete or musician, who owes nothing to an intervention, is more likely to develop and rely on his individuality and a commitment to excellence than, say, the affirmative-action college professor. With his fate largely in his own hands, his energies must go toward the development of excellence even if there is racism in the world where he functions. And, in what must be an exasperating paradox to redemptive liberals and to the grievance elite, it is precisely in areas where there are *no* interventions that blacks thrive most, both individually and *as a group.*

The black college professor, however, who is whispered to be an affirmative-action baby even by his liberal colleagues, is to a degree a child of preferential interventions, and therefore also of the opportunistic bargaining between liberals and grievance elites. He senses his fate to be at least as contingent on this politics as on meeting a standard of excellence (of which this politics tells him to be suspicious). Undergirded by an apparatus of race-focused interventions, his group identity takes on an enormous importance. Group politics frame his environment, his worldview, his career, his ideas about fairness. He mistrusts the idea of independent individuality, equates it with selfishness, iconoclasm, and danger. His view is that individuals thrive only by an intricate and unseen interdependence on the group, and that only the arrogant deny this. He thinks of individuality as permissible only within the perceived self-interest of the group. And this self-interest is *always* grounded in the contingency of black fate on the will of American institutions to redeem themselves through interventions.

One of the worst aspects of interventionism is that it forces all who "benefit" from it, and all who support it, to understand race and social responsibility primarily through received ideas. Interventions, after all, violate democratic and even moral principles in the name of the "good." There have to be justifications for this, and they have to be taken on faith. Living inside received ideas is the quid pro quo of interventionism. The black who gets a preference and the white who feels redeemed by supporting it must both unquestioningly subscribe to an entire network of ideas and assumptions that cover over the ends-justifying-the-means bargain they have made. Group preferences just "even the playing field," "help fairness along," "emphasize inclusion," "open the way to the qualified," "never lower standards," "use race as only one variable," and the like. So, both the affirmative-action professor and the white university president who supports him— one finding an illusion of equality and the other an illusion of redemption—are obligated to an unexamined orthodoxy of received justifications. Neither one can risk an open mind.

Despite recent talk of a black intellectual renaissance, it has to be conceded that few writers or thinkers in this university-based generation of black public intellectuals are on a par with Ralph Ellison, Richard Wright, and James Baldwin in his prime, or E. Franklin Frazier, Kenneth Clark, and John Hope Franklin to name only a few who came of age in the era of segregation. (Today Thomas Sowell in political and social thought and Charles Johnson and Toni Morrison in fiction and criticism are some exceptions who come to mind, though among these only Morrison is ever mentioned as part of a renaissance.) There are no doubt many reasons for this decline, and it is also true that broader America has not exactly flourished in this regard over the past few decades.

But Ellison, Wright, and the others worked out of an idea of the artist and thinker as an individual, as someone *responsible* for calling things as he or she saw them. This did not mean that they were "antigroup." (No black writer in the history of letters has understood the group culture of American blacks better or celebrated it more profoundly than Ralph Ellison.) What it did mean, however, was that understanding the group culture—its ways and its wisdom—was a part of the individuation process, a part of the responsibility of becoming an individual. Yet, having done this, these earlier thinkers shaped a vision that was their own, and offered it up despite group reaction.

Baldwin's novel *Go Tell It on the Mountain* is a penetrating study of racial shame as an ambivalent force in black life, destructive on the one hand and yet a motivation on the other—one that helped fuel the Great Migration. This is a book that would clearly outrage many blacks if it were first being published today. Ellison's *Invisible Man* was controversial with many blacks when it was published in the mid-fifties and remains so today, as do some of Wright's novels for their unflattering portrayals of certain aspects of black life. E. Franklin Frazier's devastating portrait of the black middle class has always been controversial. Zora Neale Hurston was iconoclastic enough to have criticized the *Brown* v. *Board of Education of Topeka* decision that marked the beginning of the end for segregation. As "public intellectuals" very much in the Western intellectual tradition, in which integrity involves a loyalty to the individual vision regardless of cost, they accepted that they would be criticized by their own group. The idea here, as Ellison put it so well, was that group consciousness would be "the gift of its individuals."

But today's public black intellectuals, tenured into worlds

of orthodox interventionism, contain their individual visions within a narrow and received idea of group self-interest. For the most part this generation of black intellectuals—Cornell West, Derrick Bell, bell hooks, Michael Eric Dyson, and several others—is monothematic: In a phrase, they "press the contingency." In their work black fate is shown to be contingent on the will of white American institutions to redeem themselves through interventionism—black suffering a contingency of white malevolence, black advancement a contingency of white benevolence. They may be quite different and even individual in the way they present this theme, but in the end it frames all they do and say. And always there is a genuflection to the extraordinary power of racism, which permeates the world of their work as a truth that is both utterly powerful and utterly unexamined. Racism as a kind of deity, an omniscience. What is never seen in their work is a celebration of the extraordinary range of possibility open to blacks today, or the reality of a democratic America in which possibility is ubiquitous even if a degree of racism continues.

When Cornell West says that "race matters" in his book of the same name, he is pressing for race to remain alive as a contingency, as a source of profitable and preferential interventions. He is not simply saying that it matters; he is *advocating* that it matter. Would he have said "race matters" back in the fifties when race still meant segregation, when there was no *profit* in it for blacks? Would he have advocated that race matters to that wretched pantheon of southern governors— George Wallace, Orval Faubus, Ross Barnett, Lester Maddox, and their ilk? They would surely have leaped to agree with him. And the civil rights leaders of that era, who screamed that race should never be allowed to matter, would have seen West as an enemy collaborator.

Redemptive liberalism made race into something that

"privileges" blacks in the way it once privileged white south-
erners. And this generation of writers wants race to matter
because the privileging now falls on their side. Race is the
idea of *opportunity* for them that freedom was for the early
civil rights leaders. Black difficulty, and the despair that sur-
rounds it, the hand-wringing, the sense that we are always a
second away from more racial tragedy—another riot, even
bleaker inner-city statistics—all this is the stuff of opportunity. (And, in the sniffing out of opportunity, these writers are
far more American than they imagine.) American institutions,
without moral authority, are habituated to paying out preferential interventions to keep up a certain profile of redemptive
concern.

It is important to understand here that American institutions, and redemptive liberals in general, like these writers
very much because they *also* want race to matter. When these
writers press the contingency, they open up the zone of
opportunism that allows institutions and liberals to bargain.
They give America the opportunity to *seem* to be fulfilling its
redemptive mandate. "Race matters" is code for this zone of
opportunism. What American institutions and liberals do not
like are blacks who do not press the contingency, because they
seem to offer no opportunity for white redemption. And,
when the price of redemption is as cheap as a little preferential interventionism for middle-class blacks, then it is safe to
say that this monothematic generation of black writers has
found its sinecure. This despite the fact that *whenever* race
matters in our public life, black fate remains in white hands.

Because this generation of black intellectuals functions inside
a received orthodoxy, it favors thought movements over individual visions. In addition to such standbys as Afro-centrism,
with its various schools; cultural nationalism; and various sub-

factions of Marxian analysis, today there is critical race theory in the law and even in medicine (critical medical theory).

Critical race theory sees racism as so irremediable in American life that preferential legal devices are the only chance blacks have at equality. One such device is "storytelling," in which blacks in trouble with the law are allowed to overlay the recital of their crimes with a narrative of black victimization—thus a pimp might be said to be entrepreneurial given the absence of "opportunity structures" in his racially constricted world. Lest one think "storytelling" is a ridiculous idea, it is important to note that the not-guilty verdict in the O. J. Simpson criminal trial is thought by many to be a victory for the storytelling idea. An overlay of historic racism was connected with the police handling of the case, so that the accused murderer could be cast as another black victim of racism. Following on the success of critical race theory in the law (every elite law school now has critical theorists on its faculty) is critical medical theory, which says that intractable racism gives blacks special medical needs that require preferential attention if they are to have equality in medical treatment.

Of course, more talented and individualistic writers such as Orlando Paterson, Randall Kennedy, Ntebari, and Glen Loury would seem to be exceptions to the monothematic climate of this generation. But for all their iconoclasm and criticism of group politics, none of them in the end stands up squarely against affirmative action—which is to say that they all stay on the safe side of the contingency. If anything, their individualistic reputations make their light touch on the contingency a far more effective endorsement of it than the full-court press of their more predictable colleagues.

This is simply an affirmative-action generation of black intellectuals. They are not entirely their own men and women.

They owe something not to freedom but to this contingency of black fate on white redemption. Their fear of challenging this contingency—a fear ever so reasoned, so understandable and pragmatic, so compassionate in a way—visits a degree of shame on them. And this is a sad thing to see because it compromises the dignity of the individual and the group in an era when such compromising ought to be over with. There are no good excuses for living off this contingency anymore, unless one has a greater faith in preferential interventionism than in blacks. And, yes, if one's support of affirmative action is based on one's belief in the continuing power of racism, this too points to a weak faith in black Americans. Racism is the worst excuse of all for living off this contingency.

Suppose America decided that black people were poor in music because of deprivations due to historical racism. Clearly their improvement in this area would be contingent on the will of white America to intervene on their behalf. Surely well-designed interventions would enable blacks to close the musical gap with whites. Imagine that in one such program a young, reluctant, and disengaged Charlie Parker is being tutored in the saxophone by a college student volunteer.

The tutor learns that Parker's father drank too much and abandoned the family, and that his mother has had an affair with a married man. Young Charlie is often late to his tutorial sessions. Secretly the tutor comes to feel that probably his real purpose is therapeutic, since the terrible circumstances of Charlie's life make it highly unlikely that he will ever be focused enough to master the complex keying system of the saxophone or learn to read music competently. The tutor says as much in a lonely, late-night call to his own father, who tells him in a supportive tone that in this kind of work the results one works for are not always the important ones. If Charlie doesn't learn the saxophone, it doesn't mean that he isn't benefiting from the attention. Also, the father says, "What pleases me is how much *you* are growing as a human being."

And Charlie smiles politely at his tutor but secretly feels that the tutor's pained attentions are evidence that he, Charlie, must be inadequate in some way. He finds it harder to pay attention during his lessons. He has also heard from many that the saxophone—a European instrument—really has little to do with who he is. He tells this to the tutor one day, after a particularly poor practice session. The tutor is sympathetic because he, too, has recently learned that it is not exactly esteem building to impose a European instrument on an African-American child.

Finally Charlie stops coming to the program. The tutor accepts this failure as inevitable. Sadly he realizes that he had been expecting it all along. But he misses Charlie, and for the first time feels a genuine anger at his racist nation, a nation that has bred such discouragement into black children. The young tutor realizes that surely Charlie could have been saved had there been a program to intervene earlier in his life. And for the first time in *his* life the tutor understands the necessity for political involvement. He redoubles his commitment to an America that works "proactively" to transform and uplift its poor, and that carries out this work with genuine respect for cultural differences.

The following fall, back in college, he tells his favorite history professor that he finally understands what "Eurocentrism" means. "Can you imagine," he says, shaking his head in disbelief at himself, "teaching saxophone to a poor black kid from Kansas City?"

Of course the true story of Charlie Parker is quite different from this. Though he did grow up poor, black, and fatherless in the depression, he also became the greatest improvisational saxophone player in the history of music. When he died far too young, at the age of thirty-five, he had already changed Western music forever. But the real Charlie Parker was not given the idea that his fate was contingent on an abstract racial politics that pretended to resolve history. None of this came between the real Charlie and the saxophone. What did intervene was an idea of musical excellence that prevailed in his black world—a standard of excellence that was enforced so absolutely that it would have seemed cruel to outsiders.

Some will argue that because Charlie Parker was a genius he would rise without interventions and against any odds. The regular guy would need the tutor. But this is that old double bind of redemptive liberalism that makes the black success the exception that proves the rule of black weakness. The fact is that thousands and thousands of black men and women have made respectable livings in the world of music before and after Parker's time. Few of them were geniuses, and none at all were brought ahead by government interventions that tried to transform them from musical helplessness. Many, like Parker himself, thrived even as segregation prevailed across the land.

But the point here is not simply that when people decide to do something they generally do it, with or without help. The point is that when interventionism becomes a faith, when it is implemented to transform people, it oppresses and defeats them instead. Why is this?

8.

An answer to this question became clearer to me one day when I was walking along a downtown street and ran into the retired doctor, the proud liberal I had met at dinner at a friend's house. Before I saw him, I heard my name come at me like a command to stop. Then I saw him rushing out the door of a coffeehouse, a capped cup of coffee in one hand and a newspaper in the other. "You can't tell me you're happy with this," he said a full ten feet before he reached me. When he got closer, I saw the headline of an article I'd already read that morning: MINORITY ENROLLMENT DROPS 19% AT STATE'S MED SCHOOLS. I knew immediately what I was in for. This was only days after passage of the infamous Proposition 209 (California Civil Rights Initiative) that ended race and gender preferences in state government. The atmosphere was still every bit as tense as it had been during the long, fractious debate over the proposition. So the doctor rattled the paper almost in my face and said that headlines like this were exactly what he'd expected. When I mentioned that 209 had passed only two days earlier and couldn't possibly explain this drop in minority enrollments, he read to me from the paper: "'. . . an atmosphere of unwelcomeness about the California system . . . ,'" it said. Then, looking from the paper back to me, "We've been debating affirmative action for two years in this state now. The UC regents wiped it out last year. How in God's name do you expect these kids to feel welcome?"

But I felt he was acting out a drama. I didn't truly believe him. He was proud of his anger and happy for me, an opponent of preferences, to witness it. He was also too eager to accept this single explanation for the drop in minority enrollment. Since the same article mentioned that minority enrollments had peaked four years earlier, a full two years before any public debate on affirmative action, an "unwelcome" atmosphere could at best have only added to a decline that was already in progress. Could it be that there was a shrinking pool of qualified minorities available for medical school? After all, the notorious gap in academic performance between blacks and whites, which had been closing in the eighties, began to widen again after 1991. Also, if the decline did have to do with the affirmative-action debate, wouldn't it make sense that fewer minorities would apply if they felt they would not get a preference? Is that the same thing as an "unwelcome" atmosphere?

I believe that the doctor—and the newspaper's headline writer—liked the "unwelcomeness" explanation because it was a contingency trigger. It implied that racism was behind the decline, and so it made the matter of minority enrollments contingent not on what minorities did but on what whites did. But in triggering contingency in this way, the doctor achieved something else: He allowed whites to keep agency over the problem. Now he could be outraged at white mismanagement of a black problem.

It is always the "white-blaming" explanation of a black problem—"an atmosphere of unwelcomeness"—that sets redemptive liberalism in motion. When whites are kept on the hook, they are given a degree of ownership over a black problem. And, again, they are likely to use this ownership to get history's monkey off *their* back; that is, to satisfy the mandate for redemption that history has imposed on them, not the mandate that history imposed on blacks to achieve equality and parity with whites. These different mandates to resolve history are also a great pressure to repeat history, to have whites take agency over black life and use it for their own ends.

Suppose for a moment that freedom-focused liberalism were the ideology of the day, and that medical schools were simply required not to discriminate against any group by asking all applicants to compete against a single standard of merit that reflected the actual demands of the medical profession. All who sought admission would be expected to make themselves competitive according to this demanding but fair standard, which would involve more than test scores and grade-point averages. And suppose, too, in the context of this classic liberalism, that we noticed a decline in minority enrollments. Well, as committed believers in individual freedom over group privilege, we would first of all have to ensure that the medical schools were not discriminating against individuals from minority groups. Were the standards fair in themselves, and were they applied uniformly to all applicants? Once having resolved this, we would have to look for other explanations for the decline. And simple objectivity would require that we look into explanations for which minorities themselves would be responsible. Maybe there was simply less interest in the medical profession among minorities; maybe opportunities in this area were being ignored; or maybe changes in the profession made it less attractive to minorities who now had other options available to them.

In any case, if freedom-focused liberalism prevailed, the doctor would have had just as much reason to be outraged at minorities as at whites when he rushed from the coffeehouse waving his paper. In freedom-focused liberalism culpability for a minority problem is not automatically assigned to the larger white society. The goal is to locate it where it actually resides. This classic liberalism does not allow historical mandates to confuse where culpability falls, allowing people to have agency over the part of the problem they can actually do something about.

In redemptive liberalism, however, culpability for minority problems is predetermined. Larger white America is always culpable, even when it has no control over the problem. In this liberalism any assignment of culpability to minorities automatically "blames the victim."

The key difference between these two liberalisms is that

freedom-focused liberalism tries to guide society by democratic principles, while redemptive liberalism tries to redeem society through displays of social virtue. This means that in the latter there is always a subterranean struggle for agency over minority problems. If whites have agency over these problems, then even the weight of them is redeeming: Even their failures credit them with good intentions. Redemptive liberalism is a formula for taking agency for minority problems away from minorities so this agency can serve the broader American redemption.

Here again redemptive liberalism replicates segregation, since both are based on a white need for agency over black life. In segregation, and certainly in slavery, whites used their agency over black life to assert their superiority. In redemptive liberalism they use it to assert their redemption. In both cases we blacks are left to meet our difficulties without full agency over them. In both cases we have to negotiate solutions to our problems through a self-absorbed white society.

But this liberalism also differs from segregation in one telling way: In it black leadership actually argues *against* the viability of black agency in order to engage white agency. Agency follows contingency. So to argue that black fate is contingent on white redemption is also to argue that black agency would be inadequate to the task of black development. Redemptive liberals and the black leadership make a dogma of white efficacy and black inadequacy. No doubt this was a common racist assumption during segregation, but the black leadership of that dark era did not actually go out and argue that it was true.

But what is agency?

It is ultimate responsibility combined with possession. You have agency over something—a life, a problem, the advancement of the race—when two things are true: You have the freedom that allows you to be responsible for it, and you accept that this responsibility belongs to you and not to someone else. A slave, for example, will certainly suffer in his life, but the very nature of his dehumanization is that he has no freedom in which to assume responsibility for his life. Nor does he truly possess his life, since he literally belongs to others. But once in freedom, suffering is the surest indication of where agency ought to be, of where responsibility ought to be accepted. In fact, what makes freedom "a burden," as Jean-Paul Sartre put it, is that it removes the obstacles to responsibility, so that if we do not accept it, we stand accused. A sin that no society easily forgives is that of suffering while refusing a responsibility that one is free to take.

We all, as individuals and groups, have an ambivalent relationship to agency because it is always a call to the will. To be free, and to have ultimate responsibility for a problem, is to be called upon to exert one's will to solve it. Agency demands that we *find* will even if we are mired in inertia. Thus one of the most common dodges in human experience is to deny agency by denying freedom. We say we are not free enough to have agency over our problem. We are not free enough for responsibility. If we can escape freedom, we can escape agency with its difficult call to will, sacrifice, effort, and risk.

Often we escape freedom by saying that others are to blame for our problem, and therefore *they* have agency over it and must exert *their* will to solve it. If black children are doing poorly in public schools, school officials will say it is because they come from single-parent homes, and because the school buildings are old, and because there are gangs in the neighborhoods, and so on. In other words, because of all these things, the officials do not have the freedom to take agency over their students' academic improvement. The single-parent families, the white flight that leaves no tax money for better buildings, the racist society that allows gangs to flourish—these are the people and forces that have the agency to solve the problem, not the school officials.

And what proves this, they say, is the fact that their will is meaningless. No effort they put out improves their students' performance. In fact the very weakness of their students confirms that the officials lack the agency to turn things around. If their will is ineffective, it is not because they don't truly exert it—expect academic rigor from their students. It is because they don't really have agency over their students' learning. This is how we all deny agency—by pleading that other agents leave us no freedom and so make our exertion of will meaningless.

And yet these black students and their teachers still suffer in freedom. The students may be poor, and their schools may not be the most modern, but they do freely come to school every day, where they meet adequately paid teachers who are free to teach them. And suffering plus freedom equals agency, so they have agency over the matter of their education. But they do not *accept* this agency, and so the academic performance of the children remains poor: To accept agency is to commit the will despite the obstacles and the risks.

But in most poor and black school districts there are simply too many ways for teachers, administrators, students, and their families to deny freedom and thus agency. The social problems of these areas themselves are used to argue against freedom and will. So, ironically, race and poverty are temptations to escape agency rather than prods to seize it. Very often those who educate poor blacks feel excused from the responsibility of high expectations and academic rigor by the very conditions that make such expectations mandatory. The problems are used to argue against their own solutions.

*   *   *

To put all this another way, there is something at work in our time that has broken the archetypal relationship between suffering and responsibility in the area of race. Suffering, whether caused by fate or injustice, has always been the most common spur to agency over one's experience. When we get to the point where suffering does not prod us to take agency over our lives, then we are cut off from the meaning of suffering—which is to be cut off from the meaning of reality. "Suffering is consciousness," says Dostoyevsky. And the worst thing is to suffer and yet be uninformed by it. But this is exactly what happens in situations like the above, where decades pass without teachers, officials, students, or their families learning from the suffering of relentlessly poor academic performance. In a sense something has kept them from the *benefit* of their suffering.

The culprit is redemptive liberalism. For the sake of its redemption, the United States has needed black suffering to be a helpless suffering, a suffering in which the exercise of our will is essentially futile. Only this kind of suffering is contingent on the *will* of white America to redeem itself. Only this kind of suffering serves the white redemptive mandate. Therefore this liberalism has developed an unforgivable practice, one that oppresses blacks far worse than contemporary racism: It keeps defining black problems as excuses rather than challenges, as reasons not to exercise the will. If blacks exercise will, their problems cease to be contingent on white will and interventions.

This is how redemptive liberalism steals the meaning of black suffering away from blacks. Of course the black leadership and liberal social scientists collaborate in this by vociferously arguing that blacks are in fact helpless specimens whose fate is contingent on white rescue. Thus we black Americans are given to believe that our suffering does not require our agency to end—that it will be ended by the agency of others. We are conditioned to feel that our suffering is a prod to *others* to exercise the will, and an excuse for us not to.

This last idea distorts everything. Amazingly, it transforms poverty itself from something to be escaped by seizing agency into something to be preserved for the entitlements it brings. It makes poverty and failure our currency. So what should spur the will suspends the will. In redemptive liberalism, where race is concerned, poverty is more directly related to entitlement than to suffering. Thus the relationship between poverty and suffering—probably the world's oldest incentive—is undermined.

Of course, the welfare system does not provide a luxurious life, but it has been a powerful system of incentives and reinforcements in which people—particularly women—were literally paid for having children out of wedlock, for failing to finish school, for not developing job skills, for not marrying, and so on. It is not at all an exaggeration to say that the welfare policies of the last thirty years—direct expressions of redemptive liberalism—*created* the black underclass in America. This class of husbandless homes, fatherless children, and healthy nonworking adults follows the incentive pattern of welfare policy perfectly. When poverty is never allowed to be a negative incentive (until recently generations of people were allowed to live endlessly on welfare without working) then suffering is not only muted (as it should be), but it also ceases to be a call to agency and will.

Redemptive liberalism has not been able to deal with black poverty and failure without *entitling* those conditions so that they inspire inertia rather than action: There ceases to be a relationship between hard work and self-interest.

When well-to-do black high school students do not compete favorably with others academically—even with those far less well off than themselves—this liberalism gives them a preference to college. The weak performance, even for privileged blacks, is an occasion for entitlement rather than for the experience of suffering that inspires. In that poor black school district mentioned above, the continuing failure of black children is not allowed to prod their families into agency and will. Rather it becomes a kind of profit margin for administrators who use it to press the contingency that brings in more money. In set-aside contracts, minority businesses are protected from the level of competition that makes others more competitive. In the corporate world, minorities are sometimes "handled" rather than challenged and developed.

The fact is that failure and suffering are natural and necessary elements of success. They go along with agency. When society takes them away from people for its own redemption, it does what the well-meaning tutor I mentioned did to the fictional Charlie Parker—it prevents people from finding responsibility: It leaves them entitled to irresponsibility.

The faith at the heart of redemptive liberalism is that interventions like group preferences, welfare, and hundreds of educational innovations are agents in themselves. The idea is that reasoned and systematic interventions like these can be fundamentally *transformative*, that they can metamorphose people from one condition to another more desirable one. Therefore redemptive liberalism believes that reasoned interventions make the world better and should be pursued as ends in themselves—as the center of a politics and the focus of a creed.

Aside from this, interventions are compelling in two other ways. The first is that they seem to apply reason to the task of transforming people. They work systematically and uniformly. They can be applied with precision and followed up on. Data can be collected and analyzed, adjustments made and followed up and then fine-tuned again. All this removes reform from the haphazard, and over time a social science can emerge.

Reason, as one of our culture's most revered instruments, brings credibility, if not a little magic, to interventionism and to redemptive liberalism generally. There is an *omniscience* in reason, a dispassion, that we associate with effectiveness and power, and these qualities accrue to the idea of the smart intervention that agents the social good. So the redemptive liberal identifies with the ingenuity that reason seems to promise, and feels that the reasoned intervention is a far more effective agent of change than the people this liberalism hopes to transform.

Dostoyevsky called these grand schemes of reason "crystal palaces," suggesting that they were doomed to fail because people were not "piano keys," and would finally insist on irrationality as a way of preserving their human independence. He saw an inherent elitism not in reason itself but in the idea of reason as an agent.

Still, the reasoned intervention is an extremely compelling idea because of a factor that has nothing to do with rationality. When redemptive liberals make interventions the agents of change over people, they avail themselves to one of the most popular formulas for power in the twentieth century.

This formula always begins in the same way: A society runs into a problem that shames it. At the turn of the century it was the inequities and backwardness of a society stuck in czarist-imposed feudalism—against the backdrop of a rapidly modernizing Western Europe—that brought shame to Russia. In Germany it was the grating shame of defeat in World War I, the specter of a great power humiliated. In the United States it was the shame of three centuries of virulent racial oppression that contradicted every principle the society supposedly stood for.

These societies then conjured ideas-of-the-good that they hoped would redeem them from the shame. Against the inequities of feudalism Russia would have a "classless society." Against its postwar lowliness Germany would have Aryan supremacy. And against the shame of American racism there would be a new "multicultural," "inclusive" "diversity." Always the idea-of-the-good contrasts the specific shame the society is dealing with. As a vision of what is redemptive for the shamed society, this idea-of-the-good has three qualities: it simplistically demarcates good from evil so that all who disagree with it are aligned with evil and against their nation's redemption; it is so vague that it imposes no serious accountability, sacrifice, or principle on those who support it; however, it always requires governmental and institutional interventions, if not new governments altogether.

This kind of "good," of course, is a recipe for power. The real goal of those who espouse it is the interventionism it demands from government—and, thus, control over the arms of government. To achieve this they use their idea-of-the-good to corrupt reason. By contrasting their "good" with society's shame, they make their mission so urgent that it justifies the suspension of principles that are normally seen as sacred. In fact the urgency is made to seem so intense that it becomes *virtuous* to set principle aside. This gives them a reason that is highly elastic, a reason that allows the alignment of justification with virtue and principle with evil. Now they have a reason of mere justification, a reason of license.

In the civil rights movement, that classic expression of freedom-focused liberalism, there was a return to principle and a refutation of justification. Democratic principle extended free-

dom to blacks, while segregation had been justified by an idea-
of-the-good—white supremacy. The civil rights struggle
wanted an America that was *disciplined* by democratic principles—principles conceived to stop the power hungry from justifying the curbing of freedom for individuals.

But when redemptive liberalism began to manufacture
ideas-of-the-good to redeem American shame, it made the
"good" into a currency of power, and justification into a legitimate means to power. So, suddenly, right here in the American
democracy, this classic formula for the corruption of power
came alive. The president of a public university could simply
say the word "diversity" and be justified in excluding people
from his university solely because of their race. He could do
this without a vote by the people or by any of their representative bodies. When a nation wants to redeem itself, it becomes
so afraid of its shame that it gives itself the license to fight it
with new corruptions. So this university president suddenly
has the power to intervene against freedom and principle, to
select and reject human beings and American citizens by race
alone—to do something evil as though it were a good.

Thus, in redemptive liberalism, interventionism is the
agent of social change because it brings power to the elite that
articulates "the good." It gives the university president a
license and a right to justify actions that would otherwise
make him despotic. Support for the intervention (affirmative
action) is the identification with "the good" that justifies his
having a power that is unencumbered by principle. Of course,
this is not to say that he is on par with a Nazi or a Central
Committee boss. It is, however, to say that interventionism
allows him to partake of the same corruption, to take easy
power by exploiting his nation's alignment of shame and
redemption.

Working that alignment, this president (standing in here

for the heads of most American institutions and redemptive liberals in general) supports racial preferences so as to have himself seem the master of black contingency (by supporting a policy that black advancement is said to be contingent on). This is the most powerful place for him to be, because he is simultaneously the uplifter of blacks and the redeemer of whites—a compounding of the "good." And all this for an idea-of-the-good as thin as "diversity."

But is there really enough power in this corruption to matter? After all, a little talk of diversity and a preferential admissions policy for minorities would hardly seem to constitute an abuse of power. But of course it is not that simple, though this is exactly the kind of minimization that redemptive liberals often use to defend against the charge that their masks of virtue conceal a corruption of power. I don't believe that this is a small or isolated or necessary corruption. I think it is pervasive enough to have transformed the American culture.

When a society becomes *conscious* of a shame—which is to say *accountable* for it—it creates a market in redemption. In the United States there are two types of ideas-of-the-good that capture this market in racial redemption: racial idealisms that try to refute the nation's shame, and ideas that show the redemptive liberal as a master of black contingency. Often a single idea will work in both categories. When diversity means group preferences, it is a racial idealism that also pretends to resolve black contingency.

It is the national shame invoked by ideas-of-the-good that gives them their extraordinary power to quell resistance and to justify the setting aside of principle. I once had a colleague who told me several times that he thought multiculturalism was a mere construct of racial politics that implied a false equivalency between American subcultures. This same colleague voted in favor of a "multicultural" literature course that asserted precisely this equivalency. Why?

Those who openly disagree with an idea-of-the-good are labeled with the precise shame the idea is trying to redeem. If my colleague had voted against this course, it would not be said that he had responsibly disagreed; it would be said, or implied, that he was hopelessly Eurocentric and quite possibly racist. Because shame is the active ingredient in ideas-of-the-good, they do not win agreement as much as force capitulation. In fact, capitulation to what is so obviously *not* reasonable is more redemptive than carefully considered support would be. One comes to multiculturalism as to any received faith, in a spirit of obeisance to the good.

So ideas-of-the-good have great power because they can attach both people and principles to the nation's historical shame. My colleague was famous in the department as a stickler for principle—a watchdog of sorts, who let very little get past him. But an idea-of-the-good like multiculturalism carried a far heavier load of shame than he was used to being confronted with in his principled stands. It was one thing to be called a stickler and quite another to be seen not only as an embodiment of the nation's historic shame but also as an obstruction to the nation's redemption.

The power in these race-related ideas-of-the-good is commensurate with the magnitude of the nation's shame over its racial history. This was the power in "racial sensitivity" as an idea-of-the-good that extracted a $176 million settlement from the Texaco Corporation around a racial discrimination suit in which discrimination was neither proved nor admitted to. Texaco had fought the suit for two years but paid the money after a mere two weeks of publicity about a small group of executives who may have used a racial epithet. Guilty of discrimination or not, it was the shame of their alleged racial insensitivity that caused Texaco to capitulate.

Being at odds with "racial sensitivity" as an idea-of-the-good *identified* Texaco with the nation's shameful racial history, *identified* it with historical discrimination so that it was, in effect, guilty whether or not current discrimination was proved. It quickly paid the money to separate itself from that shameful history more than from any discrimination it may have committed. This was also the power that made the Republican-dominated Congress back away from its own legislation to end race and gender preferences (ideas-of-the-good that pretend to resolve black contingency). So ideas-of-the-good may be as thin as "racial sensitivity" and as clearly antidemocratic as group preferences, but they invoke the full magnitude of America's historical shame, which then redounds to the wielder of these ideas as power.

Since the civil rights movement sent the United States on its redemptive mission, the process of using ideas-of-the-good as a means to power has become so formulaic that it now reaches far beyond race. These ideas can easily be built with themes from the American shame and redemption story: oppression, racism, sexism, Eurocentricism, domination, rigidity, logic, standards, and the like on the shame side, and liberation, relativism, leniency, compassion, feeling, diversity, and the rest on the redemptive side. Since the sixties, one has had only to construct an idea-of-the-good out of this polarity of themes in order to have power. So it could be said that this power was literally lying around like money on the ground, for anybody to pick up.

For example, in education an idea-of-the-good called "constructivism" became pervasive in elementary education by saying that children arrived at, or constructed, answers to questions and problems in their own way. This relativistic idea tends to make the process of arriving at an answer more important than the correctness of the answer, and encourages teachers to prejudge the "developmental readiness" of students so as not to ask too much of them. The damage done to educational standards and expectations by this idea is immeasurable. But it has thrived because it sought to liberate children from the repressive uniformity of rigid, answer-oriented (and by implication male-dominated, left brain, Eurocentric) expectations with an appreciation that children have individual, cultural, and gender-specific learning styles that must be validated.

"Constructivism" would never have appeared before the sixties, before the redemptive mandate that gave America a new polarity of shame and redemption. This is an idea-of-the-good that literally tries to separate the act of teaching from the shameful dominance and oppressiveness of the American past. It makes teaching into a redemptive struggle: The teacher is not a teacher but a missionary, and the student is not a student but a victim in need of liberation and solicitude. (As always, one must be helpless for the other to be redeemed.) Because constructivism is aligned with "the good," with the sloughing off of oppressive uniformity, it shames all who would argue against it; it associates them with oppressiveness. Thus it extinguishes resistance and licenses those who support it to bend traditional standards and principles. The power in this idea-of-the-good is what brought us "wholistic reading," "naturalistic reading," writing for content rather than correctness, and many other "innovations."

Again, this kind of power lies around like money on the ground. Black college students can pick it up with an identity-centered idea-of-the-good and demand more black faculty, or a black cultural center, or a black student newspaper. Johnny Cochran picked it up with an idea of racial justice (as opposed to criminal justice) against the shame of police corruption and used it to stunning effect in the O. J. Simpson trial. The excesses of feminism have come from this power, from perfectionistic ideas of women's justice taken to absurd levels of license. Corporate America, long indifferent to racial politics, has been brought by this power to a near obsession with diversity. In fact this was the power that changed the framework of social reform from freedom-focused liberalism to redemptive liberalism. In many ways the latter was conceived as a framework out of which ideas-of-the-good could extract the power latent in America's shame.

Every time this power is used—every time, for example—the talk turns seriously to "diversity"—it is used *against* the principles and standards on which a free society depends. Principles, like a fairness based on individual rights or advancement by merit, are terms that help us work for a better world within the integrity of freedom. After all, when excellence decides who gets in, race and gender do not. The fact that some do not have the same chance to develop excellence is *not* an argument against excellence, nor is the problem of such people solved by denouncing standards. *In fact a fair standard of excellence is what both clarifies their problem and points to its solution.*

But ideas-of-the-good have nothing to do with principle or freedom or the problems of people who don't meet standards of excellence: They are always about power. And this is what "diversity" shares with "the classless society" and the "Aryan nation." If America's little ideas-of-the-good are not comparable in scope to these infamous twentieth-century tragedies, it is only because our democratic principles and our commitment to freedom have protected us from ourselves. We are fortunate to wrestle with our shame and our ideas-of-the-good within a society that still treasures freedom over the "good."

So the great American democracy will probably survive "diversity." The damage done is pervasive (because it is a damage to principles), but other societies have survived even more entrenched ideas-of-the-good.

Of course, the group that will suffer the most from "diversity"—and so many other mellifluous ideas-of-the-good—is black Americans. Group preferences, multiculturalism, welfare rights, and diversity all picture a world in which blacks have great power to shame institutions into exceptionalism but little or no agency to transform themselves into full equality. In redemptive liberalism, black power moves others to action; it tries to win action on behalf of blacks in the form of interventions, but it is not a power that individual blacks can use to transform themselves: It is a power that attaches to ideas-of-the-good rather than to people.

Since the famous *Bakke* decision in 1978, in which the Supreme Court reaffirmed affirmative action by allowing race to be a consideration in medical school admissions, Asian-American students at UC Berkeley moved from 12 percent of the undergraduate student body to 31 percent in 1996. In that same period blacks moved from 3.9 to 5.1 percent. Black students were the beneficiaries of one of the most aggressive racial preferences in the nation. (Blacks were admitted with SAT scores averaging almost 300 points below those of Asian-Americans.) Conversely, this same aggressive use of preferences meant that the number of Asian-American students was suppressed during this period. So, as they were being openly discriminated against, Asian-Americans gained 19 percentage points while blacks, who benefited from intense discrimination in their favor, gained only 1.2 percentage points.

Many factors no doubt contribute to a discrepancy like this. But an extremely important one is that Asian-American students clearly put more faith in their own agency, while blacks tend to have more faith in such ideas-of-the-good as group preferences, identity-support programs, the presence on campus of black role models, and so on. As if to reinforce this pattern, the university itself assumes an agency over black admissions (and even the general happiness of black students on campus) that it never assumes for Asian-Americans. Then it loudly proclaims its own virtue for doing so. As their own agents, Asian-American students present themselves well prepared and in large numbers for admissions. Blacks, on the other hand, with an elaborate idea-of-the-good (preferences) behind them, and an entire politics of identity to support their move into university life, come in smaller numbers, are less prepared, maintain lower grade-point averages while in college, and drop out in higher numbers than all other groups.

Ideas-of-the-good, which bring such a convenient power to institutions and to so many race professionals and researchers, are now the worst enemy of blacks because they defeat us psychologically and spiritually. These ideas do not respond to black need; they respond to American shame. They treat the nation's shame, not the people whose suffering shames the nation. But they also go farther than this: They facilitate the most self-destructive theme in black life.

When a people finally emerges from an oppression, often the most tragic effects of the experience are never discussed or fully named because of the pain and shame they carry. Others also do not want them named because of the shame that redounds to them as well. So what is rarely spoken about in relation to American slavery is the extent to which this "peculiar institution" suppressed the human will and the sense of agency in those it enslaved. American slavery was a crucible that used physical and psychological terrorism to make the slave feel that his independent will was a futility and his agency an illusion. His wife could be had before his eyes, his family sold away for cash. For more than three centuries black American culture evolved inside this crucible where the exercise of our will was acceptable only when it served the self-interest of others. And, because independent will and agency are what make us human, blacks were made to live with a hopelessness and terror around our full humanity, and to find a degraded security only in a diminished humanity— not a humanity utterly without will, but a humiliated humanity in which will could only be exercised through a subterfuge or from behind a mask. The Louis Armstrong smile was such a subterfuge, the cover of a driven man.

But the crucible of slavery suppressed agency and will in another way. Think for a moment of the irony of the slave who worked hard, who showed initiative and imagination. This sort of slave—and there were many—probably worked out of the natural human need to *use* the life God had given him, despite the absurd predicament he found himself in. And yet this *using* of his life only fattened the people and the political system that oppressed him. Thus he was made to live against himself. His own will—his own full humanity—betrayed him, making him a collaborator in his own oppression.

So, at the very heart of black American culture is an ambivalence toward individual will and agency *when applied outside the black community*—where, historically, independent black will either put blacks at great risk or made them collaborators in their own oppression. The tragedy here is in the way this history has injured the connection between the individual will and *group* self-interest. Black culture associates self-interest with the collective will in the form of protest because protest is what won our freedom. But once in freedom, individual will is the best—if not the only—way to pursue self-interest for individuals *and* for the group. Therefore, in this context, it is not too much to say that our history ill suited us for freedom by surrounding the very idea of individual and independent will in so much ambivalence.

And this may be the most significant difference between those irrepressible Asian-American students whose numbers constantly rise at UC Berkeley and the black students whose numbers never improve much even after decades of affirmative action. Maybe to an extreme degree among the Asian-American students, the exercise of individual will is tied to both personal *and* group self-interest. The individual's performance honors both him- or herself and the group. And this honor is earned through performance; it is not bestowed before performance to enhance self-esteem in the *hope* that improved performance will follow. The reward of honor is an incentive to performance, not to self-esteem or group pride or group identity. And this priority has nothing to do with stoic individualism or lonely effort. Asian-American college students are well known for working together in small groups, and this capacity for cooperation no doubt contributes to their academic success. But even group effort has no meaning until the individual assumes agency and exerts will over his or her own performance. Individuals who reject agency and will are understood to be of little value to the group, and honor is withheld from them.

Educational reform for black students is most often the reverse of this. Its focus is generally on identity enhancement and self-esteem. Its profound mistake is to assume that performance follows self-esteem, when in fact it is the other way around: Performance follows high expectations; not high self-esteem. The proof of this is that Asian-American students routinely test lower on self-esteem measures than blacks, who routinely test higher than any other student group. Also, *whenever* academic expectations are raised for black students, their performance improves. I would argue that this is because only high expectations engage the will of black students; give them agency over their own performance; and reward them with a true and indisputable honor when they succeed. Schools that are successful at educating young blacks, like the Frederick Douglass School in Harlem or the Piny Woods School in Mississippi or Xavier University in New Orleans or Marva Collins school in Chicago, make their students the agents of their own performance and expect them to exercise their own will in achieving excellence.

These schools do what most (not all) Asian cultures do in the United States: They tie individual and group honor to performance (as blacks do in music, entertainment, and sports). No doubt these schools also support and encourage their students to perform, but support is not construed to mean being excused from high expectations. Only high expectations counter the natural human ambivalence toward the will. And when that ambivalence has been deepened by the conditioning of three centuries of oppression, then high expectations are the *only* antidote.

Of course, the redemptive liberal will say that his idea-of-the-good does not preclude high expectations for black students. He will say, for example, that we can have racial preferences and high expectations at the same time. But this is, in fact, an impossibility. As an idea-of-the-good, racial preferences take agency over the black difficulty and promise to substitute for the will of blacks by engineering them into equality whether they will it themselves or not. They always lower expectations, ask less of blacks than of others, excuse us from full competition, and so on. They whisper that we will "get in" anyway.

In a much-talked-about speech on racial matters at the University of California at San Diego in 1997, President Clinton said that the opponents of preferential affirmative action had nothing to put in its place. Thus, by default, group preferences should remain. Of course, being asked to support a policy by mere default is never very persuasive, nevertheless this argument was very revealing. The default argument suggests that the president's first concern is simply that *some* policy be in place—a concern that makes it seem as though he believes *only* in ideas-of-the-good as agents of black advancement. Nowhere in that speech, or elsewhere, did the president say that black Americans must be the agents of their *own* fate. He seemed to say that if not affirmative action, then "we whites" must come up with some other government-sponsored idea-of-the-good because "our" national decency (redemption) is tied to how "we" execute "our" agency over black problems.

The president's idea of justice is white agency over black problems. Black agency is of no real use to him because it makes blacks and not him the master of contingency. When justice is white agency, blacks have to be seen as inadequate in the name of justice. So the president projects black inadequacy not because he believes blacks are inferior, but because he is in competition with blacks for agency over their problems.

The redemptive liberal has high expectations only for whites, and his subtext is always a message of white power and black impotence.

In his San Diego speech the president bemoaned the 81 percent drop in black enrollment at the University of California Boalt Hall law school after racial preferences were banned in California as if it were a *white* shame. Because he was blind to even the idea of black strength, he did not consider that a drop in black enrollment of this magnitude was first of all a *black* shame. After all, most of the blacks who apply to this select law school come from middle-class and advantaged backgrounds. Most come from select undergraduate programs. Isn't it permissible for the president to say to blacks that a performance this terrible is unexplained by disadvantage and is therefore unacceptable? (Simple logic would suggest that the certainty of a preferential admission to law school may have depressed black performance more than any other factor, even as the certainty of a nonpreferential admission may have prodded whites and Asians to higher performance.) The president of the United States did not call on black strength to solve this black problem, however. He called on white strength. But then, the redemptive liberal always lowers expectations for blacks as a "way" to black advancement.

What makes race-related ideas-of-the-good so alluring is that they represent a vision of the "good" that goes beyond the mere freedom that democracy provides. They represent a vision of America perfected of its shames—a vision of *anti-shame.* "Diversity" and "multiculturalism" have no substance as ideas except that they connote perfection exactly where America was shameful. Their appeal is their promise of transcendence, their vision of a society that not only compensates for historical shame but that also compensates for the unpredictability of freedom itself through social engineering. To be identified with such ideas is to be associated with an enterprise far greater than the laissez-faire possibilities of freedom: It is to be associated with something morally beautiful.

It is not surprising that the baby-boomer generation and its first presidency should be so given to ideas-of-the-good. It fell to this generation to deal with the great sense of shame that came into American life after the civil rights victories of the mid-sixties and during and after the Vietnam War. Transcendence of the shameful racial past, the foreign adventurism, and the "homogenized" "gray flannel" culture of "commercialism" at home became literally a generational mission. Thus, in both the personal and the political, we became obsessed with ideas-of-the-good that connoted perfectibility and transcendence.

We are the generation that created heretofore unknown categories of these ideas. In the category of psychology alone there must be a hundred such ideas around the themes of self-esteem, self-realization, quality time, personal space, personal empowerment, and so on. And now in the category of "political correctness" there are entire subcategories of ideas around group preferences, race and gender identity, cultural differences, diversity, and *so* on. Our generation came of age in the certainty that our parents' generation was hypocritical, repressed, and duplicitous, not to mention racist, sexist, and imperialist. In order to "make the world a better place," we chose "the good" over freedom.

Freedom has too many disadvantages for a generation bent on redeeming a great shame. In freedom, principle always trumps any idea-of-the-good. *Whenever* an exception is made to this, we are giving something else a greater importance than the freedom the principle is trying to ensure. If we set merit aside to bring in more blacks or women, then we are saying that the presence of blacks and women is more important than the freedom from race and gender bias that the principle of merit is there to enforce. We are saying that their engineered presence is more important than their freedom to be present or not. And we say this because we need the moral symbolism of their presence more than we need freedom.

Shame gave the United States the need for a "good" that was transcendent and beautiful, and in so doing, it left us with a virtuousness that is the enemy of both freedom and black self-determination.

9.

When blacks won the great civil rights victories of the six-ties, it confirmed that we were *right* about the injustice we had endured. The effect of this was to bring the idea of jus-tice alive as it had never been before—for us and for the United States generally. Suddenly justice was a hope and a faith. A quick look into black culture shows that, for obvious reasons, blacks had never been much convinced of social or historical justice. We were a blues people, and much of our art and music consoled us for the intractable injustice of our fate.

But then, here was a great victory for the idea of social jus-tice, and with it came the temptation to think of justice as power in itself. If it could bring freedom, could it now also become a currency in freedom? Black American culture was changed by this victory for justice. We stopped being a blues people reconciling ourselves to injustice and became instead carriers of justice. Wider America saw us this way too, and, as such, we became the carriers of America's redemption.

A serious problem with justice achieved is that it can bestow a feeling of entitlement that forever outruns what it can deliver. In our minds justice works on the principle of equivalency: We feel entitled to positive prospects somewhere in proportion to the injustice we endured. For blacks this obviously opens up an expansive sense of expectancy and desire—so much so that it may have introduced the theme of entitlement into the group identity. In any case this is how justice became a pressure to put our fate into the hands of other people, to believe in race-related ideas-of-the-good, and to see interventionism as the best agent of our advancement. Justice gave us a dangerous idealism.

The *logic* of justice—that racial history can be resolved by pursuing a level of entitlement today in proportion to yester-day's injustice—has the appeal of something complete, like music. And there are few blacks who have not heard this music. But after justice is asserted and freedom is won, justice does little to build up those that injustice kept down. It doesn't teach, impart skills, explain the etiquette of networking or the intricacies of high finance. In fact, people can languish in jus-tice just as surely as in injustice. The difference is that, having won justice, they will hear its music more powerfully. And instead of moving beyond the injustices of history, they will be drawn back into them.

To contemplate equivalencies to past wrongs, one has to be forever gazing backward into history, forever assigning value to past injustice, forever reacquainting oneself with the horror of it. Justice's trick is to keep the feeling of injustice alive.

This aspect of justice brings a very specific problem to racial reform, one that might be called, a little awkwardly, *ulterior-ality*. This occurs when the ulterior goals of a given policy or reform are far more important than its announced ones. In a sense I have been talking about this problem throughout this essay. Most of our racial policy in America suffers from ulteri-orality because our announced goals—"diversity," "inclu-sion," "multiculturalism," and the like—are only a pretext for unstated and more powerful ulterior goals—pursuing redemp-tion, reclaiming moral authority, appearing the master of con-tingency, seeking racial monopolies, and winning the look of equality.

In racial reform, from the Great Society to the morass of affirmative-action programs and policies, the only true accountability has been on the ulterior level. These reforms are almost never accountable to the goals they actually announce. If, for example, corporations and universities truly cared about affirmative-action goals, they would insist on quotas in order to be more accountable to their preferences. But of course the true accountability of these institutions is to their ulterior goal of showing themselves committed to racial justice—a goal that can be satisfied almost entirely through symbolism.

Ulteriority has given American racial policy a decidedly "postmodern" feel. Its subtext is more important than its text, and we read this subtext for its representation of justice. We read for justice, and approve of policies that seem to redress the imbalance of historical injustice. Thus virtually any policy that seems to *give* something to poor blacks is acceptable, even when it is a conspicuously ineffective policy. Redemptive liberals have been against welfare reform for the past thirty-five years, even though welfare policy has clearly been an incentive to teenage pregnancy—the mechanism that literally grows the underclass. If this form of welfare was blatantly ineffective, it seemed to the redemptive liberal—in its very license—fair and just, given America's racial history.

Because the subtext of our racial policy is linked to justice, and therefore also to social virtue and the historical mandates of white redemption and black equality, we all have powerful reasons for supporting these policies, with little regard for their actual effectiveness or their destructive side effects. This is similar to, though broader than, the "Glazer trap" that I mentioned earlier. The heavy load of ulteriority carried by our racial policies—their ulterior symbolism of justice, social virtue, redemption, and the like—causes us to support them *against reason* as iconographic representations of all we want to be identified with.

If the United States actually required racial policy to be effective in terms of its announced goals, then it would surely fail on the ulterior level. To satisfy ulterior goals, this policy cannot demand high levels of performance from blacks without seeming cruel and unjust. Effective racial policy would necessarily "blame the victim" and therefore be unjust on the ulterior level. In other words, racial policy that effectively advances blacks would ask more than give, and so would not seem *justly* to resolve history or redeem America. Policy that excuses blacks from standards (thereby engendering weakness) meets the ulterior requirement of symbolic justice and so seems redemptive for America. The ineffectual policy is the virtuous policy; the policy that works is too "mean" to be just.

Ulteriority has ruined the enterprises of redemptive liberalism—the Great Society, affirmative action, multiculturalism—just as its absence made Roosevelt's New Deal rather effective. Roosevelt was not trying to redeem the nation from a terrible moral shame. He was not trying to exhibit a justice that would resolve history. Unlike the redemptive liberal, he actually wanted what he said he wanted—to stimulate the country into an economic recovery. There was no vast network of ulterior goals that was more important than the goals he announced. Therefore his programs and policies were not nearly as iconographic. People did not support them *against reason* because of their symbolism. If a policy was ineffective, it could be abandoned without the anxiety that an underlying moral mission was also being abandoned.

The simple absence of ulteriority gave New Deal liberals two crucial advantages over today's redemptive liberals. First, because they were accountable primarily to effectiveness, they had the flexibility to find what was effective—thus the famous experimentalism of the New Deal era, in which many programs and policies were tried. By contrast, policies like affirmative action are so laden with ulteriority that they become all but intractable, taking years of torturous debate to alter in any way.

A second great advantage was that New Deal liberals could make demands on the people they were trying to help. They could have high expectations of people, could ask them to work harder, to conquer their fears, and to take more responsibility. They could make it clear that there was no magic— that ultimately it was up to Americans to agent their own recovery and advancement. In redemptive liberalism the government must be the agent because *it* has to be seen as redeeming the nation's sins. Demands are not made on those being helped; rather they are *excused* even from what is expected of others. And in the inertia that this condescension often lulls them into, they become something like a permanent redemptive opportunity for the larger society—exploited once for labor and now again for innocence.

Ulteriority mires every public policy it touches in duplicity. It forces all who support such policies into the deception that they truly care about the policy's announced goals when, in fact, their true commitment is to unacknowledged needs. In most public policy there is a territory of unacknowledged need that competes with the policy itself for accountability. But in the area of race, this territory is so vast that virtually no race-based public policy will ever have enough surface accountability to be effective in advancing minorities.

Instead such policy will always be essentially iconographic—emblematic of a subterranean negotiation of needs and with no real expectation of surface effectiveness. It will be widely accepted (rather than supported), even as it is rather obviously ineffective, because people will sense that it is iconographic, that it represents a negotiation of needs they are unable to talk openly about. And when people can't talk about a social problem, policy around that problem has the added burden of serving as an occasion for negotiating all that cannot be said. (The affirmative-action debate bares the burden of countless ulterior debates: How racist is America thirty-five years after the civil rights victories? Have blacks overplayed their "victim" card? and so on.) The problem is that when the real action is ulterior, the policy itself is neutered.

The ulteriority of America's racial reform opens the way for an extremely profitable corruption that I will call *indirection*.

When most of the accountability for racial policies is ulterior, so that they never have to produce results, people can put forward policies, theories, and programs that can stay around for decades without ever having to demonstrate effectiveness. This opening has made black problems into very profitable opportunities.

If you can give a black problem a racial cause, you can turn the business of solving it into a monopoly for blacks. To do this you will use an *indirecting* idea—an idea that directs us away from the human cause of the problem by automatically assigning it a racial cause. In the field of education racial self-esteem and identity have been extremely effective indirecting ideas. (These have been discussed earlier as contingency triggers. The difference is that with indirection the emphasis is more on racial essentialism than on racial woundedness as the element that brings white obligation. Indirection tries to describe a racialist world in which race is *the* all-important variable whether or not there is racism. It obligates society to racial reform even when there is no racism to redress.) Racial identity, particularly, opens the way to *exclusive* racial territories, where whites are pushed away and blacks drawn in. Suddenly only black people can deal with a problem (inner-city police work, for example) because it is presumed that their race gives them the unique sensibilities to deal with it.

Skin color carries a group essentialism that America must accommodate.

So indirection directs us away from the *human* sources of social problems by giving them an immutable racial character that, in turn, qualifies *only* blacks (or the appropriate minority or gender) to work on the problem. Indirection is the mechanism that provides affirmative-action supporters with an endless supply of rationalizations. It suggests that there are pedagogical and cultural reasons (having little or nothing to do with racism) for people of a given race to attend to their own kind. So, if this results in an employment monopoly for that race, then so be it. It is for the good.

Indirection is a kind of racial smoke-and-mirrors that lets people create racial monopolies in the name of racial essentialism so that we end up with a virtual racial unionism in public schools, universities, and other institutions. Under its auspices several tracks of employment and business have opened up exclusively for minorities—human resource departments in corporations, student services in universities, Black Studies Departments, Hispanic Studies Departments, Education Departments in universities, "minority beats" in journalism, the diversity-training business, inner-city social work and police work, racially oriented consulting businesses, and countless social and educational programs: racial essentialism as employment opportunity.

But the greater problem in all this comes from combining the racial exclusivity of indirection with the lack of accountability to announced goals that comes from ulteriority. Not only do minorities have an exclusive lock on dealing with, say, an inner-city school district, but they also are not accountable for succeeding. The larger society is funding the school district (federal as well as local money) and even allowing minorities to monopolize the jobs in it (by giving the nod to the racial essentialism of role-modeling) in order to satisfy its ulterior need for redemption. And, having satisfied this need, it does not bother to enforce accountability in the schools and programs it funds: Ulteriority opens the space for the corruptions of indirection.

When problems that minorities suffer are defined in racial terms, they become vulnerable to this lethal combination of ulteriority and indirection. The general rule is that indirection follows the money. No one bothers to use indirection where there is no possibility of funding. Inner-city schools, where there is a steady funding source (even if more limited than in suburban schools) have been devastated by years of indirection. Hundreds of educational innovations have been brought in on the supposition that they are somehow right for minority children—everything from bilingualism to holistic reading to Afro-centrism to all-male academies. "Innovators" learned to tailor their innovations to minority children because that was where the funding was. Yet outside school there is less widespread "innovation" going on in how to shut down crack houses, curb gang activities, or slow the rate of teen pregnancy. These problems have fewer steady funding sources than school systems and therefore are rarely indirected. They are less shrouded in the racial specialness that gives minorities a monopoly. They rarely enter the matrix of indirection and ulteriority, and are instead left to a few earnest and underfunded citizens' groups to handle as best they can.

However, minority academics have successfully indirected the study of their cultures within universities, not only creating employment monopolies but also getting universities to fund separate Ethnic Studies Departments. These departments and programs study nothing that cannot be studied in already existing departments, but ulteriority makes them immune from this kind of evaluation. Universities win their redemptive profile by merely funding them. Rarely can the faculty and administrations of the best U.S. universities say what it is that their Ethnic Studies Departments actually study. Nor can they point to an academic methodology that might justify their status as academic disciplines. And because universities rarely hold these departments accountable to academic excellence, they are not likely ever to achieve serious academic stature. They exist for two reasons: because universities agreed to fund them, and because minorities then made them objects of indirection.

When a problem is defined as racial and it slips into the matrix of ulteriority and indirection, it becomes a minority monopoly without accountability: It becomes a business.

10.

Whenever we self-consciously *use* race (or ethnicity)—whether out of hate or love—as a tool, a convenience, a proxy for disadvantage, a currency of entitlement, a means to power, a basis for group preferences, then we are using it precisely to gain the license to break the normal human and democratic principles we live by for some ulterior reason.

Because our truest motivations for using race are always ulterior, every race-based public policy or program—from segregation to affirmative action—is a duplicity in which what we say merely rationalizes goals that we are unable and unwilling to state. Today there are many who feel that a little duplicity in the service of a good cause is "progressive." What they miss is that whenever we make a race-based policy, we also make rationalization the carrier of social virtue in our society.

And this has given the redemptive liberal an anxiety, a self-doubt, that was never present in the freedom-focused liberals of the civil rights era. On some level redemptive liberals know that their program of social virtue is supported by mere rationalization, that there is no honor in the means to their "good." They look like conservatives looked in the civil rights era when they stood by as people *rationalized* segregation as a "good." Today the score is even. Both the Left and the Right have let race seduce them into trying to have rationalization fly as virtue.

No society has tried and failed in more ways than America to make race serve the "good."

The only way out of this situation, and the corruptions of indirection and ulteriorality, is to use a strictly *human* analysis of our social problems, even when those problems are caused by race. This means simply seeing those who suffer social problems as first of all human beings and American citizens, so that whatever the source of their problems may be, their needs are understood to be human and not racial. The United States should now be racially experienced enough to understand that a multiracial democracy simply cannot have an obligation to meet the racial needs of its citizens; its only obligation can be to address their human needs *without regard to race.*

Even the victim of blatant racial discrimination is a human being who suffers because his race has been used to stifle his humanity. His society owes him the aggressive enforcement of antidiscrimination laws, but in the name of his humanity and citizenship, not in the name of his race. Where many races live together, race always threatens to be an enemy of humanity. Racial identities are erected and maintained for purposes of aggression, defense, and power in relation to other races. Then, of course, cultures may attach to them, but a multiracial/multicultural democracy is obligated only to humanity, citizenship, and the democratic traditions that provide the basis for a common national culture. Because race is atavistic—people only inherit it—it is inherently antidemocratic and must always be kept apart from entitlement, privilege, and power. A healthy democracy is always at war with race.

What about racial pride? When I look back, as a black, at the many achievements of blacks in history, I am especially proud because these human beings achieved so much despite the fact that their race was relentlessly used to dehumanize them. My *racial* pride honors the extraordinary and tenacious *humanity* of black Americans. It was our humanity that prevailed despite our race, and it prevailed precisely because we refused to become *merely* our race. Race doesn't struggle; people do. And the story of black America is one of the greatest stories of human irrepressibility in world history.

So our post-sixties embrace of race as the centerpiece of our group identity and politics is very possibly the saddest chapter in our struggle. It suggests a self-destructive confusion, as though we ourselves were completing the mission of our former oppressors, finally accepting what they had tried for three centuries to convince us of—that there is an essentialism in race that justifies the pursuit of a racial power. Of course, redemptive liberalism helped this confusion along by actually giving out entitlements by race. Nevertheless, for the past thirty years we black Americans have fought ferociously for racial power and have tried to make race a permanent part of social policy. We have strained to believe exactly what we survived by refusing to believe—that race is a legitimate means to power, that there is real deliverance in race. This is sad because no group in this country has struggled harder to earn its humanity despite its race than blacks. To now insist so fiercely on racial recognition is to sell our birthright for a pot of porridge. There are no resources in race; there are only human resources.

*   *   *

American social policy should always be conceived to stimulate the human resources of those it seeks to help, and to discourage the illusion that race can substitute for these resources. I think those who make this policy should always keep race *out* of their proposed solutions even, and especially, when race contributes to the problem. Whenever race becomes part of the solution, symbolic justice becomes the true goal. Race should be avoided even in the analysis of problems because it will only make the problem *responsible* to history. Soon people will see the possibility of power in it. Ideas of indirection—racial identity, self-image, cultural pride, role-modeling, and the like—will cast a racial spell. Experts will emerge. Pools of grant money will materialize. And the problem itself will be beyond help. The very moment race enters the analysis, corruption not only follows as night follows day, but the problem is then celebrated as a mark of society's shame. This is what makes it more profitable to sustain than to solve.

Problems that race usually obscures—inner-city unemployment, poor academic performance among black children, high college dropout rates, and so on—become easier to work on when defined in strictly human terms. This not only takes the profit out of them, but it also suggests ways that institutions can *stimulate* their solution.

Without race we realize that these problems are not suffered by helpless "others" but by free and responsible citizens. We are no longer trying to redeem the United States or resolve our tortuous racial history; we are dealing with poor reading skills or a lack of preparedness for employment or too many pregnant teenage girls made that way by too many young men.

Whatever reform we devise must start from the same fact that life itself starts from: To be human is to be responsible. Correspondingly, living without responsibility constitutes a kind of inferiority, even when people are prevented by oppression from carrying responsibility for themselves. This was the kind of inferiority imposed on blacks by slavery, on Jews by Hitler. When welfare or affirmative action robs people of full responsibility, they also impose inferiority. What social reform in the future must nurture is a hunger for responsibility.

Along with this, it must think of equality primarily in terms of competitiveness. There is no full equality for any group that is not educationally and economically competitive. But this goal rests on another one: a devotion to excellence. And, to use Booker T. Washington's phrase, this should be a "drop-our-buckets-where-we-are" kind of excellence. This does not mean an "accommodationist" excellence that excuses the larger society from any obligation to help us out. (Remember, race and all its grudges are out of our calculations.) It means instead that we internalize a devotion to academic and economic excellence that *is not contingent on any assistance we might or might not get from the larger society.*

American inner cities have the poorest facilities for basketball in the country, yet produce the greatest basketball players in the world. This is because the people there have internalized a commitment to excellence in basketball that supersedes what others do for them. The same no-excuses, drop-our-buckets-where-we-are commitment to excellence has to be adopted toward academic performance. So every effort at social reform has to start from this understanding, and communicate that without the internalization of these principles—responsibility, competitiveness, excellence—nothing anybody does will ever do more than feed a vast petite bourgeoisie of social-problem workers.

There are, of course, no devices that will magically instill these principles in people. They have to be asked for and expected. Today faith-based reforms are much talked about because they have been more effective than most approaches with seemingly intractable inner-city problems. But I think this is because God, not race, is called on. And God, as we know, not only has high expectations but also offers the incentive of his grace to those who take on difficult principles. No one but God can offer this, but his formula of high expectations and incentives can be useful to less divine entities.

Incentives reverse the psychology of contingency, making the fate of the beneficiary contingent on what *he* does rather than on what *others* do. Put another way, incentives reverse the entire psychology of redemptive liberalism in which we rush to help people before they help themselves. It does not preclude help, but it gets before it gives, which of course is how reality works. Incentives require people to be accountable to themselves. In this way they separate black Americans from the idea of their helplessness, and commit them instead to their own agency.

I am what is called a black conservative because I came to feel that redemptive liberalism was the second American betrayal of black freedom. The first was oppression; the second delusion.

But am I lonely?

In the United States today the presumed loneliness of the black conservative is an article of liberal faith. Our public discussion of race is still framed by redemptive liberalism. It is still "correct" to consider black helplessness and victimization the "truest" black experiences. The BC exclaims that opportunity is as available to blacks as a change of mind—that awareness of opportunity is the same thing as opportunity. So the black conservative's loneliness is a liberal propaganda against black opportunity.

In thinking about all this I am reminded of a passage by the Italian writer and Holocaust survivor, the late Primo Levi, in which he describes what it was like to be liberated from the concentration camps. He makes the point that there was not much happiness in liberation, that "almost always it coincided with a phase of anguish." He says of those liberated, "Just as they were again becoming men, that is, *responsible* [emphasis added], the sorrows of men returned." For our purpose here, the important idea is not the reference to sorrow but the equation of humanity with the word "responsible." Liberation did not bestow happiness: It bestowed *agency*.

And with agency came the responsibility to create opportunities for survival. In those bereft, postliberation circumstances, agency must have felt like yet another cruelty. But it was precisely this almost impossible responsibility for their own survival that restored humanity to the survivors.

Perhaps the greatest corruption of redemptive liberalism was that it made opportunity seem to be a happiness that could be delivered by others in redress for past suffering. This magical view of opportunity flattered the liberal with the illusion of his own moral power. It seduced blacks into the delusion that opportunity was the same thing as liberal interventions, and that these interventions were justified by past suffering. Opportunity became distorted into a magic that resolved history, into something given out of obligation and received as redress.

In fact enduring opportunity grows out of the struggles of agency. When a person or group truly takes agency over their own fate, opportunity materializes virtually out of their concentration—especially in a free society. Others can be helpful, but only if they never take agency for problems away from those who suffer them. It is easy to help those who are the agents of their own fate, and impossible to help those who are not.

Redemptive liberalism has some of the qualities of what Albert Camus called "metaphysical rebellion"—"a claim, motivated by the concept of a complete unity, against the suffering of life and death and a protest against the human condition both for its incompleteness, thanks to death, and its wastefulness, thanks to evil." The metaphysical rebel is mad at God as the "the father" of death and suffering and hopes to give the world an order that God has failed to give it. He has ideas-of-the-good that imply perfectibility.

But suppose we contrast metaphysical rebellion with an older and more universal practice: the initiation. Here the focus is not on an argument with God, nor is it on an idea-of-the-good that will deliver "unity" and perfectibility. All initiations are submissions to reality that carry an expectation of agency in which individual responsibility is seen as a community's best chance against suffering, evil, and premature death.

In this sense liberation was initiation for the survivors Levi describes above. And the civil rights victories were an initiation for black Americans—a confrontation with a new reality that carried an expectation of agency. The metaphysical rebellion of redemptive liberalism—the need to redeem America by offering minorities a "unity" against their suffering—*interrupted* the initiation of black Americans into greater freedom. It relieved us of agency just as new freedom asked us for new levels of individual responsibility.

I think social/racial reform in the United States is best thought of as an initiation into a free society where so much depends on individual agency. The idea of initiation spares

Americans the need to resolve history or win redemption or find "unities." Reform can be practical. How do people become professionals? How are businesses built? How does money flow in a free society? What beliefs are self-defeating? How much and what kind of effort is actually required to achieve certain goals? Where must the individual be assertive? Where is cooperation crucial?

Reform must be an initiation into the demanding responsibilities of freedom, not a faith in ideas-of-the-good.

There are four elements to reform as initiation:

1. There must be an insistence that agency be assumed by those who suffer.
2. There must be no magic around the idea of opportunity. Opportunity is not a happiness in itself; it is responsibility and imagination applied to reality.
3. Race should *never* play a role in social reform for many reasons, not least of which is that it is *always* used to help people avoid full agency for their fate. It always transforms the responsibility that free minorities should carry into a commodity that others will use for their own moral power. Race absolutely corrupts those who use it for redemption and absolutely weakens those who use it for advancement.
4. All social reform that hopes to initiate minorities fully into society should be based first and foremost on high expectations. The subliminal message of a high expectation is "You *can* and you *must.*" *The high expectation is the only credible assertion of equality that a society can make.* It not only shows faith in the human equality of minorities but it also holds them accountable for demonstrating that equality through performance. Other reforms make equality a suppo-

sition; high expectations demand its manifestation. I do not believe that minorities will ever have true respect for a reform that does not demand as much or more from them as from others. The most dehumanizing and defeating thing that can be done to black Americans, for example, is to lower a standard in the name of their race. Here the black is asked to accept the inferiority he resisted for centuries, to identify with it as proof of his victimization, to hold it, and to use it. But in the high expectation there is a faith in his equal humanity, intelligence, and skill. And when he meets that expectation, his equality becomes unassailable.

But redemptive liberalism was also the way many whites avoided an initiation. This liberalism was based on the faith that America could essentially "fix" its history by engineering away the effects of historic racism. So at its heart there is an equation between "fixing" and innocence, and a view of fate as something very malleable.

But, in fact, fate is as irrevocable as history. We don't "fix" it because it is who we are. We can only try to make the best of it.

The civil rights movement should have initiated Americans into a healthy fatalism—the recognition that we cannot "fix" all the bad things we do as a way of reestablishing our lost innocence. If we can now understand this, we might also understand the importance of our democratic and human principles. Only the *discipline* of these principles—fairness by individual rights, opportunity by freedom, advancement by merit, equality by individual rather than by group, and so on—can keep us from doing "bad things" in the first place.

Only this discipline can separate race from power, desire, and evil.

A mature society—with an awareness of the irrevocability of what it does—understands that the unforeseen consequences of "fixing" always make new problems. (When we were going to "fix" history by essentially making welfare a "right," we created an underclass we now live with as a new irrevocable part of history.) Of course there is no magic or perfection in democratic principles either. But they do at least demand a level of individual responsibility that makes us far less vulnerable to racism and to all the schemes to "fix" its damage.

In American life race will always be an opportunity for evil. It will also never enter public life except to excuse people from the responsibilities of democracy. Only the separation of race and ethnicity from public entitlement and an insistence on the freedom of the individual will curb the evil and license that always follow race into public life. Call it conservatism, classic liberalism, or simply a hard-earned wisdom. But race can never be a pretext for activism; it has to be a pretext for discipline.

WRESTLING WITH STIGMA

1.

I have a white friend who has told me many times that he feels no racial guilt despite the fact that he was raised in the Deep South before the end of segregation. Though he grew up amid the inequality and moral duplicity of segregation, and inevitably benefited from it as a white, he says simply that he did not invent the institution. He experienced it as a fate he was born into. And when segregation was finally challenged in the civil rights era, any solidarity that he felt with other southern whites was grounded more in a sense of pathos than in any resistance to change. So, he says, there is no "objective basis" for racial guilt on his part.

Recently I was surprised to hear the novelist William Styron, a southerner by birth and upbringing, say on television that he, too, felt "no [white] guilt" despite the fact that his grandmother had owned slaves as a girl. And there was something emphatic, even challenging in his pronouncement that discouraged questioning. For as long as I can remember, I have heard white Americans of every background make this pronouncement.

This is certainly understandable. White guilt threatens the credibility of everything whites say and do regarding race.

Specifically it threatens them with what I have called ulterior-ality—the suspicion that their racial stands do not come from their announced motivations but from ulterior ones driven by guilt. We can say, for example, that the white liberal bends over backward because he is motivated by guilt even though *he* says he is motivated by true concern. Or we can say the anger of the "angry white male" is simply his way of denying guilt. We can use guilt to discredit *every* position whites take on racial matters. So it is not surprising to hear so many reflexive denials. When people like my friend or Styron do this, they are disclaiming ulterior motives. They want us to accept that they mean exactly what they say.

But I, for one, very rarely do accept this, or at least not without a glimpse past their words to the matter of ulterior motive. This is because there simply is no social issue in American life more driven by ulterior forces than race. One reason for this is that white American motivation in racial matters has gone largely unexamined, except to attribute support for policies like affirmative action to white goodwill and nonsupport to white racism. "White guilt" is almost a generic term referring to any ulterior white motivation. But the degree of ulteriority in American race relations is far too great to be explained entirely by guilt. I think the great unac-knowledged event of the civil rights era was that white Americans became a stigmatized group. I also believe that our entire national culture of racial and social reform—the poli-cies, programs, norms, and protocols by which we address race-related problems—has been shaped more by the stigma-tization of whites than by any other factor, including the actual needs of blacks.

Ironically, it was the idea of equality that brought stigma to whites. In the civil rights era, when white America finally accepted a legal equality that would extend to different races, it also accepted an idea that shamed it. For three centuries white America had used race to defeat equality. It had indulged in self-serving notions of white supremacy, had transgressed the highest principles of the democracy, and had enforced inequality on others while possessing the ideas to know better. The American racial shame is special in that slavery and segregation were knowing indulgences. The nation's first president had denounced the institution of slavery and freed his own slaves, yet it would take two more centuries for segregation to be outlawed. An evil strung out over the centuries and conducted in a full knowledge of itself.

America's new commitment to equality in the civil rights era brought with it an accountability for all this. What no one could have foreseen was that a great shaming of white Americans and American institutions was a *condition* of greater racial equality. In a sense the new embrace of equality floated the nation's racial shame, unanchored it, so that it rose to the surface of American life as a truth that the nation would have to answer for. As a result equality in the United States has depended on a vigilance that associates this racial shame with whites and American institutions. This association, of course, is the basis of white stigmatization.

In this way the idea of equality has established a social framework in which white Americans are no longer "universal" people or "Everyman" Americans. Today there is a consciousness that whites are a specific people, a group with a history, a fate, and a stigma like other groups. So far equality has worked by bringing whites down into stigma rather than by lifting blacks and other minorities up out of it. The morality implied in equality stigmatized whites as racist and thus gave them a group identity that they are accountable to in the eyes of others even if they reject its terms. Very often the strongest group identities

form in response to stigmatization because stigmas are a kind of
fate, a shared and inescapable experience. In any case the history
of white racism, the idea of equality, the stigma created by these
two things, and the need to wrestle with this stigma as the way
back to decency—all this gave white Americans a new post-
sixties identity that was *not* universal. In the way that blacks had
been stigmatized as inferior, whites, too, became a group marked
by a human incompleteness.

As black Americans know only too well, to be stigmatized
is to be drawn into a Sisyphean struggle for redemption from
the accusation carried by the stigma. It is also to lose some of
one's freedom to the judgment, opinion, or prejudice of others. White Americans now know what it is like to be presumed
racist and to have that presumption count as fact against
them. What blacks know is that one group's stigma is another
group's power. Stigmatized as inferior, blacks were deprived
by whites of freedom itself. Now stigmatized as racists, whites
can easily be extorted by blacks for countless concessions. So,
when a group fights against its stigma, it is also fighting for its
freedom from the power of another group.

Being white in America has always meant being free from
racial stigma, as if "whiteness" might be defined as simply the
absence of stigma. Until recently we never had stigmatizing
epithets for whites of any real power. "Honky" hardly com-
pared to the visceral "nigger." (Today the term "racist" is
quite effective against whites, but this is a post-sixties phe-
nomenon.) This *absence* of stigma was always the blessing of
being white in the United States, while color, even "one
drop," was a stigma in itself that defined all who carried it as
alienated "others." In America whites have been the "it," not
the "other," so they have always had a rather myopic view of
race as essentially a problem of "others."

One of the bestselling books on race during my youth was a book called *Black Like Me* by a white man, John Howard Griffin, who had chemically darkened his skin and traveled the South passing for black. What made the book sensational was that a white man had volunteered for the black stigma, the experience of the alienated other. But it was little more than a novelty book that put off many blacks because its very premise tended to mistake the black stigma for the entire black experience. The reader, whom the narrator presumed to be white, was invited to watch one of his or her own in the land of the "other." And the black "other" was shown to endure little melodramas of man's inhumanity to man at which the "good" white reader could be appropriately aghast. This began an age when white America was invited not to *see* black life but to be aghast at it. However, the book's greater sin was to suggest that even if whites were morally obligated to support equality, race was still a problem that affected others.

But equality finally gave whites their own racial otherness. The idea of democratic equality—explicitly applied beyond even the boundary of race by the 1964 Civil Rights Bill— showed white Americans *as a group* to have betrayed the nation's best democratic principles. Even though it was the white embrace of these principles that brought the civil rights victories, it was the *need* to embrace them in the middle of the twentieth century that proved the white betrayal of them. And this profoundly injured the legitimacy of whites as a group in relation to principles of any kind. They had used race to give themselves license from principle.

Of course, the fact of a group finding a pretext for violating its own principles was hardly new. What is new is for an oppressive group to embrace equality at the expense of its own moral legitimacy, so that it has to live with those it once oppressed without the moral authority to enforce the society's best principles. This situation, this fate, comprises the "otherness" of white America today. It is alienating to live with this stigmatic association with shame, and to have lost standing in relation to the principles one was raised to cherish, to watch the institutions of one's society—from the family to the public schools—weaken for want of demanding principles, and to be without the necessary authority to restore them, to lose "universality," to have one's angry former victim define social morality, to feel both a little guilty and falsely accused, to feel pressured toward a fashionable relativism as toward racial decency itself—all this and more has come to whites as an experience of "otherness" that I believe is the unexamined source of U.S. racial policy since the sixties. The idea of racial equality has given a new and unique contour to the white American experience. Perhaps a *White Like Me* is now called for, a book that looks into the world behind the white stigma and reports back to us.

One point such a book would no doubt make is that stigmas are often double binds. The stigma of whites as racists *mandates* that they redeem the nation from its racist history but then weakens their authority to enforce the very democratic principles that true redemption would require. And this is no small problem because the United States is no better than its principles. It may be the first country in the world to have principles and ideas for an identity.

The promise of the American democracy was that freedom, and the discipline of principles that supports it, would be the salvation of humanity. This discipline would replace the atavistic power of divine kings and feudalism with a power grounded in reason. Principle would be not only the soul of America, it would be the basis of its very legitimacy as a nation among nations. The principles of freedom were the case for a new nation.

And yet race is always an atavistic source of power, going back to a primordial source, back to the natural order. Like a divine or natural right, it comes from God or nature and presumes that one's race is free to dominate other races by an authority beyond reason. The white racist believes that God made whites superior, so that even a democracy grounded in principle and reason is not obligated to include blacks and other races. Atavistic power always oppresses because it is immune to reason and principle. The great ambition of democracy was precisely to *free* man from atavistic power through a discipline of principle that would forbid it.

I say all this to make the point that white racism was no small thing. It was a primitivism, a return to atavistic power and, most important, a flaunting of the precept that America was founded on: that the freedom of man depended on a discipline of fragile and abstract ideas and principles. White racism made America illegitimate by its own terms, not a new nation after all, but an "old world" nation that used God as an excuse for its oppression and exploitation, a pretender to reason and civilization.

So, what happens today when a white American leader, even of the stature and popular appeal of a Ronald Reagan, questions affirmative action on grounds of principle? The Reagan administration, famous for its disbelief in racial preferences, refused to challenge these policies because even this extremely popular president lacked the moral authority *as a white* to enforce the nation's very best principles—advancement by merit, a single standard of excellence, individual rather than group rights, and the rest. Not only have white Americans been stigmatized as betrayers of principle, but those principles themselves have been stigmatized by their association with white duplicity.

Here were whites exclaiming the sacredness of individual

rights while they used the atavism of race to deny those rights to blacks. They celebrated merit as the most egalitarian form of advancement, yet made sure that no amount of merit would enable blacks to advance. Therefore these principles themselves came to be seen as part of the machinery of white supremacy, as instruments of duplicity that whites could use to "exclude" blacks. The terrible effect of this was the demonization of America's best principles *as they applied to racial reform.*

This situation, I believe, has given post-sixties racial reform its most stunning irony: Because difficult principles are themselves stigmatized as the demonic instruments of racism, white Americans and American institutions have had to *betray* the nation's best principles in racial reform in order to win back their own moral authority. For some thirty years now white redemption has required setting aside the very discipline of principles that has elsewhere made America great.

If not principles, then what? The answer in a word is deference. Stigmatized as racist, whites and American institutions have no moral authority over the problems they try to solve through race-related reform. They cannot address a problem like inner-city poverty by saying that government assistance will only follow a show of such timeless American principles as self-reliance, hard work, moral responsibility, sacrifice, and initiative—all now stigmatized as demonic principles that "blame the victims" and cruelly deny the helplessness imposed on them by a heritage of oppression. Instead their racial reform must replace principle with deference. It must show white American authority deferring to the nation's racial tragedy out of remorse. And this remorse must be seen to supersede commitment to principles. In fact, any preoccupation with principles can only be read as a failure of remorse. "Caring," "compassion," "feeling," and "empathy" must be seen to displace principles in public policy around race.

But deference should not be read as an abdication of white American authority to black American authority. American institutions do not let blacks, in the name of their oppressive history, walk in the front door and set policy. It is important to remember that these institutions are trying to redeem their authority, not abdicate it. Their motivation is to fend off the stigma that weakens their moral authority. So deference is first of all in the interest of white moral authority, not black uplift. Certainly there may be genuine remorse behind it, but the deference itself serves only the moral authority of American institutions.

And this deference is always a *grant of license*—relief from the sacrifice, struggle, responsibility, and morality of those demanding principles that healthy communities entirely depend on. And virtually all race-related reform since the sixties has been defined by deference. This reform never raises expectations for blacks with true accountability, never requires that they actually develop *as Americans,* and absolutely never *blames* blacks when they don't develop. It always asks *less* of blacks and exempts them from the expectations, standards, principles, and challenges that are considered demanding but necessary for the development of competence and character in others. Deferential reform—everything from welfare to affirmative action to multiculturalism—is the license to be spared the rigors of development. And at its heart is a faith in an odd sort of magic—that the license that excuses people from development is the best thing *for* their development.

Nowhere in the ancient or modern world—except in the most banal utopian writing—is there the idea that people will become self-sufficient if they are given a lifetime income that is slightly better than subsistence with no requirement either to work or to educate themselves. Nowhere is there the idea that young girls should be subsidized for having children out of wedlock, with more money for more children. And yet this is precisely the form of welfare that came out of the sixties—welfare as a *license* not to develop. Out of deference this policy literally set up incentives that all but mandated inner-city inertia, that destroyed the normal human relationship to work and family, and that turned the values of hard work, sacrifice, and delayed gratification into a fool's game.

Deferential policies transform black difficulties into excuses for license. The deferential policy maker looks at the black teen pregnancy problem with remorse because this is what puts him on the path to redemption. But this same remorse leads him to be satisfied by his own capacity to feel empathy, rather than by the teenage girl's achievement of a higher moral standard. So he sets up a nice center for new mothers at her high school, thereby advertising to other girls that they too will be supported—and therefore *licensed*—in having babies of their own. Soon this center is full, and in the continuing spirit of remorse, he solicits funds to expand the facility: *It was not joblessness that bred the black underclass; it was thirty-five years of deference.*

Deferential policies have also injured the most privileged generation of black Americans in history. Black students from families with incomes above seventy thousand dollars a year score lower on the SAT than white students from families with incomes of less than ten thousand a year. When the University of California was forced to drop race-based affirmative action, a study was done to see if a needs-based policy would bring in a similar number of blacks. What they quickly discovered is that the needs-based approach only brought in more high-achieving but poor whites and Asians. In other words, the top quartile of black American students—often from two-parent families with six-figure incomes and private-school educations—is frequently not competitive with whites and Asians even from lower quartiles. But it is precisely this top quartile of black students that has been most aggressively pursued for the last thirty years with affirmative-action preferences. Infusing the atmosphere of their education from early childhood is not the idea that they will have to steel themselves to face stiff competition but that they will receive a racial preference, that mediocrity will win for them what only excellence wins for others.

Out of deference, elite universities have offered the license *not* to compete to the most privileged segment of black youth, precisely the segment that has no excuse for not competing. Affirmative action is protectionism for the best and brightest from black America. And because blacks are given spaces they have not won by competition, whites and especially Asians have had to compete all the harder for their spots. So we end up with the effect we always get with deferential reforms: an incentive to black weakness relative to others. Educators who adamantly support affirmative action—the very institutionalization of low expectations—profess confusion about the performance gap between privileged blacks and others. And they profess this confusion even as they make a moral mission of handing out the rewards of excellence for mediocre black performance.

A welfare of license for the poor and an affirmative action of license for the best and brightest—the perfect incentives for inertia in the former and mediocrity in the latter.

But this should not be surprising. Because "racial problems" have been a pretext for looking at blacks rather than at whites, we have missed the fact that most racial reforms were conceived as deferential opportunities for whites rather than as developmental opportunities for blacks.

*   *   *

Because deference is a grant of license to set aside demanding principles, it opens the door to the same atavistic powers— race, ethnicity, and gender—that caused oppression in the first place. Again, the United States was founded on the insight that freedom required atavisms to be contained by a discipline of principles. The doctrine that separates church from state is an example. And race, ethnicity, and gender are like religion in that they arise from a different authority than the state. They come from fate, or some would say from God, and so are antithetical to democracy, which comes from an agreement among men to live by a social contract in which no single race can be validated without diminishing all others.

But thirty-some years of deferential social policies that work by relieving us of principle have joined atavisms to the state as valid sources of power. (This also happened recently in Eastern Europe, where the unifying principles of Communism collapsed so that the atavisms of tribe, clan, and religion surged back as valid sources of power and entitlement. War has been the all too frequent result.) A quick look at America's campuses reveals what I have elsewhere called a "new sovereignty," in which each minority carves out a sovereign territory and identity based on the atavisms of race, ethnicity, and gender. And this new atavistic sovereignty supersedes the nation's sovereignty and flaunts its democratic principles. One is a black or a woman before one is an American.

It is no accident that preferential affirmative action became the model for racial and social reform after America's great loss of moral authority in the sixties. Affirmative action is an atavistic model of reform that legalizes the use of atavisms in place of principles right in the middle of a democracy. In this way it mimics the infamous Jim Crow laws that also legalized the atavism of race over democratic principles. In Jim Crow, white supremacy was the motivation; in affirmative action it was deference. The first indulgence in atavisms so wiped out white moral authority that it made the second indulgence inevitable.

To take all this a step further, liberal whites and American institutions also shifted the locus of social virtue itself from principles to atavisms. Since the sixties, social virtuousness has lost its connection to difficult and raceless principles and become little more than a fashionable tolerance for atavisms. Of course tolerance of different races, ethnicities, and genders *is* virtuous. But moving out of a spirit of deference, white liberals and American institutions have asked that these atavisms be tolerated as legalized currencies of power. This is how the virtue of tolerance becomes a corruption of democratic fairness—you don't merely accept people of different races; you *validate* their race or ethnicity as a currency of power and entitlement over others.

This is the perversion of social virtue that gave us a multiculturalism that has nothing to do with culture. The goal of America's highly politicized multiculturalism is to create an atavistic form of citizenship—a citizenship of preferential status in which race, ethnicity, and gender are linked to historic victimization to justify entitlements unavailable to other citizens. Culture is a pretext, a cover. The trick of this multiculturalism is to pass off atavisms as if they were culture. So people think they are being "tolerant" of "cultural diversity" when, in fact, they are supporting pure racial power.

In fact multiculturalism actually *suppresses* America's rich cultural variety, because much actual culture does not mesh with victimization. A troublesome implication of jazz, for example, is that blacks are irrepressible since they created one of the world's great art forms in the midst of oppression. It is images of helplessness that highlight their racial atavism as a source of entitlement. So the black cultural genius for self-invention and improvisation that made jazz possible are not drawn out and celebrated in multiculturalism. Nor are the many other cultural ingenuities—psychological, social, and political—by which blacks managed to live fully human lives despite their hard fate. Culture gets in the way of multiculturalism.

But multiculturalism is the kind of thing that happens when a democracy loses the moral authority to protect the individual citizen as the only inviolate unit of rights. In any society atavisms can only be repressed, never entirely extinguished. They are always waiting for the opportunity to wedge themselves into the life of society under some high-sounding and urgent guise. No one invents the moral mask better than those driven to have their race, ethnicity, or gender bring them a preference over others—whether white segregationists or minority supporters of affirmative action. And when the majority of a society is stigmatized for past betrayal of principles, and when those principles themselves are emblems of duplicity, then primitive atavisms easily present themselves as salvation itself. Multiculturalism masks a bid for pure atavistic power; it is an assault on democracy that Americans entertain because they feel they must. It was conceived not to spread culture but to win some of the territory opened up by the weakened moral authority of American institutions.

2.

The conspicuous extravagance of President Johnson's Great Society, the drama and scope of its almost wild assault on poverty, was not primarily about ending poverty. Of course, no one would have objected had poverty been conquered. But

I don't believe the *will* to conquer it was what gave the Great Society its frenetic energy. This came from a kind of denial—a reflexive insistence that the United States was *not* the shameful country that the civil rights victories had shown it to be. The Great Society *screamed* that the stigma of whites was not true, that they were actually a fair and compassionate people who would now "end poverty in our time." But this grandiosity was primarily a measure of the shame that stigmatization had delivered.

The Great Society was America's first rather hysterical wrestling with racial stigma. It may have involved an abundance of good intentions, but its ulterior motivation of fending off stigma turned it into a hyperbolic, contrived, and ineffective exhibition of racial and social virtuousness.

I worked in four separate Great Society educational programs in the late sixties and early seventies, and they were all very exciting, though no small part of this excitement was the fact that we didn't really know what we were doing. Our mission was simply to be "innovative," but this only meant rejecting the traditional ways of doing things, whether that way made sense or not. (I believe that the Great Society helped launch the trend of wanton educational "innovation" that so injured American schools in the seventies and eighties.) The trick to "innovation" was simply to stigmatize the traditional way of doing things with the shames of America's past—racism, repression, intolerance, rigidity, exclusion, "mechanized" learning, "rote learning" (that is, repression), and so on. Against these heavy stigmas, any idea that was deferential to the "oppression" of students—whether racial, ethnic, patriarchal, or simply the result of a repressive and mechanized society—and that licensed students to a relief from traditional expectations, was "innovative" *and* socially virtuous. By the magic of this formula we could think of ourselves as socially committed, innovative educators, even though the majority of us were teaching without any training and many among us had only a few more years of education than did our students. We needed a formula.

And so did the Great Society. It, too, did not know what it was doing and so needed a formula by which it could seem innovative and virtuous *without having to accomplish anything.* From its inception the Great Society was defined by ulteriorality: It was infinitely more accountable to its ulterior goal of fending off a shameful stigma than to its announced goal of "ending poverty in our time." So it created a chimera of exciting good works through the magic of deference and license. Ask less, excuse from principle, stigmatize tradition, mock the difficult struggle for mastery. And at the end of the day everyone could claim "at least" to have been *well-intentioned* even if nothing had been accomplished. And this claim was entirely the point of everything, because the true "war" of the Great Society was not against poverty: It was against stigma.

Because the Great Society was largely stigma-driven, it gave America its first clear example of what I will call *iconographic* racial reform—reform that exists for what it represents rather than for what it does. Iconographic programs and policies function as icons of the high and honorable motivations that people want credit for when they support these reforms. And this representation of high motivation is the true reason for their existence. The announced goals of these programs and policies will be very grand, the better to represent their high virtuousness, yet vague so that their inevitable failures will not be held against them. (Today any program with "diversity" as a goal is an example.) Supporters of iconographic policies are primarily concerned that these policies function as icons of their high motivations, not whether they achieve anything or whether they mire those they claim to help in terrible unintended consequences. Societies go to this kind of policy when they need an iconography with which to fend off stigmatization. So there is always an inverse relationship between stigma and icon. They will be literally two sides of the same coin.

In the area of race, iconographic policies are based on deference and license because these themes give whites and American institutions the imagery with which to dissociate from the stigma that says they are racist and oppressive. Virtually all American institutions of any size, public or private, have "diversity programs," regardless of whether they achieve anything or whether they are even constitutional. Iconographically they represent a dissociation from America's historical shame that wins the institution at least a *look* of moral authority. We say, "At least they are sensitive to the problem" (America's racism), and so we give them credit for decency. This is the credit they need in order to do business in a shamed society.

Iconographic reform is facilitated by a specific corruption that has been at the heart of racial reform since the sixties: *the separation of social virtue from principles.* Stigmatization pressured Americans into a racial virtue that would be easy to exhibit—a virtue unencumbered by principles. Aside from the fact that difficult principles were themselves stigmatized with white duplicity, they were also verboten because they were not deferential and did not offer license. Principles make demands on people, ask for sacrifice, require delayed gratification—impose difficult struggles. Whites and American institutions not only lacked the moral authority to base racial reform on demanding principles, they have also had deferentially to offer blacks the license to sidestep them. Principle had simply become a barrier to racial virtuousness for white America.

And yet principles are what bring social policy to bear on human character. In fact good social policy invariably applies difficult principles to a social problem. Even the famous "make-work" programs of the New Deal reinforced rather than set aside the entire constellation of principles around work. The postwar GI Bill was an *incentive* to initiative and individual responsibility that helped veterans pay for college *after* military service and *after* they won admission in open competition. In these policies social virtuousness was achieved through principles that engaged the characters of those being helped by requiring responsibility, discipline, effort, and so on. But then the conceivers of these policies were not so stigmatized that they needed to cut principle away from virtue and subvert policies into deference and license in order to claim their own decency.

Since white America has had *openly* to carry the shame of the nation's racial past, only an iconographic racial virtue has been forthcoming.

To improve the performance of black college students, for example, universities would have to (1) refuse to defer to the victim-focused black identity (without denying historical victimization) that alienates and demoralizes young people who have experienced very little racial victimization, and that weakens their connection to the principles that high performance requires; (2) not grant the license of a preference that denies black students the competition with whites and Asians that excellence absolutely requires; (3) not demonize the very principles—rigorous intellectual effort, skill mastery, grade and test performance—by which those who compete with black students are strengthened.

Yet we regularly hear university officials defend racial preferences by dismissing the achievement that comes from subscribing to demanding principles. They constantly argue that high grade-point averages and test scores "should be only one measure," that they render up only a "narrow range" of student—and in the expression of a common liberal racism—that "they would make UCLA all Asian." There is a serious move now in California to eliminate the SAT exam altogether as a university entrance requirement. Iconographic reforms like preferences, which are determined to treat blacks only with deference and license, are always supported by a denigration of the very principles that blacks most need to develop.

Universities, like most American institutions, have learned to separate their racial virtuousness from principles. Of all the university presidents across the country who claim to be concerned about low black student performance, no more than the smallest handful would dare stand up and utter the above three points. One way out of this squeeze is to make the obvious point that better elementary schools would help black performance greatly. But there is no disagreement on this anywhere; it is an easy stigma-free assertion of principle that skirts the matter of university responsibility. Even if early education for blacks were vastly improved, these three points would still be necessary on the university level for black students to perform well. But virtue and principle are joined in them. They make principle the *means* to virtue. Thus they are verboten because they bring down stigma rather than fend it off. And it is stigma more than black students that these universities care about.

The need of American institutions to have racial virtuousness without the stigma-risk of principles is what has mired America in iconographic racial reform. And because this reform answers white stigma before all else, it has an indifference, if not resistance, to the true needs of blacks that mimics the indifference of oppression. Diversity, multiculturalism, affirmative action, and the propriety of political correctness are all icons of white racial virtuousness that never engage the independent will, character, or determination of blacks. With deference and license they try to buy white moral authority. And in these iconographic schemes, blacks themselves are often mere icons, carriers of white virtuousness, brought in to "diversify" an environment. They are as humanly invisible to the purveyors of diversity as they were to the segregationists of old.

A law professor says, "I want blacks in my classroom when I teach constitutional law. The diversity of opinion helps us better understand the Constitution." But are blacks human beings or teaching tools? Is it good for human beings to be made to play this role, to be brought in, often in defiance of the standards, because their color is presumed to carry a point of view that diversifies classroom comment? And doesn't this transform even those blacks who win their place purely by merit into factotums of racial sentiment? In flight from stigma, the pressure is to see blacks themselves only in iconographic form.

Iconographic reform has not flourished by serving whites alone. It has also served a symbiosis of black and white elites— on the one hand liberal and moderate whites, government, corporate, and institutional leaders, and on the other hand minority leaders from civil rights organizations, education, government, social services, unions, and politics. I refer to this latter group as the *grievance elite*, not because its members are mired in anger and a sense of grievance but because their political power and moral authority derive almost entirely from their group's racial grievance against the United States.

Today the grievance elite is not only those highly visible minority leaders whom we all know; it is also quite a vast second tier of people who believe both that the group's grievance should translate into iconographic reforms, and that the group's advancement is *dependent* on this parlay of grievance into reforms. Anyone from affirmative-action college professors to minority journalists on the minority "beat" to corporate "human resource" administrators, bilingual educators, minority student counselors, diversity trainers, and so on clearly *qualifies* to be a member of the grievance elite. But it is really this *idea* of historic grievance turning into entitlement in today's world—and for people with little or no experience of discrimination—that joins one to this elite. Some who inescapably benefit from iconographic reform reject this idea. Others who benefit not at all base their very hope on it.

Since the sixties the grievance elite, like the liberal elite it bargains with, has been an establishment in its own right, which means that self-preservation has been its first goal. Just as the liberal establishment (which includes virtually all American institutions) needs an iconography to exhibit its racial virtuousness, the grievance elite has to exhibit the *power* to transform the grievance into entitlements. This elite does not risk its self-perpetuation on developmental reform, where its power would actually be measured by how well it developed blacks' competitiveness with others. Neither elite, liberal or grievance, risks its immediate needs—the look of racial virtuousness and powerfulness, respectively—by asking for enduring black development. Thus, since the sixties, blacks have been led primarily by people who so lack faith in them that they have been unwilling to risk their own fortunes on higher black performance. None of the major groups most devoted to advancing blacks into the mainstream of American life has been willing to ask that higher black performance in education and economic development be the first and most relentless method of reform—*whether or not other reforms and resources come from the larger society.*

So the grievance elite is essentially a reactionary elite—timid, conservative, and cunning enough to tie its own fate more to what it can gain from threatening institutions with stigmatization than from helping blacks become more competitive. In fact, it gains *more* from black failure than from black success because the liberal elite responds only to black failure—which always threatens it with stigma. No group likes to recite statistics of inner-city collapse more than does the grievance elite, which lives by the credo that black difficulty is *always* a grievance that justifies an entitlement.

The same iconographic reform that allows the liberal elite to look virtuous allows the grievance elite to look powerful. It is the deference and license at the heart of this reform that enables it to serve the ulterior needs of both elites at the same time. To offer deference is virtuous; to be deferred to indicates power. And the grievance elite, like the liberal elite, is more accountable to this ulterior goal than to any announced goal. As historic victims blacks are not at risk of being stigmatized as racists. However, they endured the shame of impotence, of living for centuries without the power to help or even to free themselves. So the grievance elite succeeds when it satisfies a deep longing in blacks to see their leaders as powerful enough to win deference and license from the former oppressor. By satisfying this ulterior and largely symbolic goal, this elite preserves itself whether or not it contributes to the development of blacks.

Wherever unannounced and ulterior goals are the truest goals that groups are driven by, the reforms they settle on will function more as icons than as reforms: The reform will have to *represent* the unannounced need. Thus the perfect reform for the liberal and grievance elites to settle on was race-based preferential treatment. A preference is both deference offered out of virtuousness and deference won out of power. It represents the most profound need that each race has in relation to the other, the need to overcome the deep shame that each represents for the other. On an iconographic level, the racial preference resolves America's terrible racial history.

So it is not surprising that most racial reform since the sixties, including much of the Great Society, has been more preferential than developmental. The preference has simply been the currency of trade between the races. Almost every entrée to the worlds of public and private employment, education, business, licensing, contracting, and even voting districts has been restructured by preferential schemes. Our new language of race—"multiculturalism," "diversity," "pluralism," "tolerance," and the like—is a euphemistic gloss that tries to divert our attention from the naked preference.

For the grievance elite itself, preferences function as *patronage*—as favors it can pass out to blacks and others as proof of the power it wields. Of course poor and working-class blacks do not get preferences (and, thus, patronage) because preferences go almost exclusively to the wealthiest and best-educated blacks. (White women from the middle and upper classes get several times more benefit from preferences than any other group.) But again, the working-class black is flattered by the preference his well-off cousin gets to the Ivy League because it stands as an icon of black power. Patronage works by reputation as well as by literal benefit. It flatters all those in the group that is eligible for it with a kind of specialness that inspires their loyalty even if they never actually receive it. This is why group preferences are the lifeblood of the grievance elite. A little preference wins a lot of loyalty.

Some argue (Christopher Caldwell of the *Weekly Standard* is one) that the National Democratic Party hopes that affirmative action will be a patronage that holds the desperate units of its coalition together. Not only does it flatter all of America's minorities, it also flatters women, the group reputed to have won the White House for the Democrats in 1992 and 1996. Clearly, corporate America has recently discovered the value of affirmative action as patronage. Hiring goals and timetables, accelerated promotions, and diversity-training programs are, as they relentlessly say, "Good for business." What they really mean, of course, is that these very inexpensive preferences buy them considerable loyalty with minorities inside and outside the corporation. They also give the company a degree of protection from discrimination suits. But most importantly, they *iconographically* dissociate the company from the racist stigma. Preferences are a kind of moral advertising by which the company says, "We are aware of America's racial shame, and are so dissociated from it that we discriminate in the opposite direction."—preferences not as developmental racial reform but as patronage that tries to buy the corporation into public favor.

Preferences work so well as iconography and patronage because they are based on magical thinking. They presume that the application of deference and license in the form of a group preference will magically resolve the inequities caused by historical racism. The idea is that preferences can magically *make* equality. If black students are admitted to UC Berkeley averaging almost three hundred points below whites on the SAT exam, then somehow the deference and license of admitting them will magically eliminate the disparity. Specifically the illusion is that deference is the same thing as development, that a preference will in itself instill three hundred points of skills and competency in the young minorities who receive it. Another assumption is that preferences will magically render the disparity in competence *insignificant* over time as other "untestable" talents inevitably emerge in minorities—a formula by which preference + time = equality.

This is the magic that redounds in greater iconographic power for preferences. This is what makes them so much more attractive to the liberal and grievance elites than the rigor and sacrifice of earnest developmental reform. Their "magic" immediately comes back to both elites as a look of greater virtuousness and greater power, respectively. Selling magic is hardly a new way of pursuing one's goals, but it is always a sign that a social movement has lost its idealism and selflessness and become an entrenched elite with privileges to preserve and bills to pay.

After it came to light that, during a surreptitiously tapped phone conversation, an executive of the Texaco Corporation had used the phrase "black jelly beans" in reference to certain black employees, the entire corporation became instantly stigmatized as racist. For two years before this disclosure Texaco had claimed innocence in a racial discrimination suit that had been filed against it. But once stigmatized, the company immediately gave up the fight and settled the suit to the tune of $176 million, $35 million of which was earmarked for diversity training. Beyond this the company agreed to a strict system of "goals and timetables" in which managers are at least partially evaluated for their promotion of blacks, a system that virtually puts a bounty on black heads. They have set up an independent commission with no fiduciary responsibility to the company to ensure that hiring and promotions are contributing to diversity. They have agreed to give money to black magazines like *Emerge* and *Black Enterprise*. Diversity training is mandatory throughout the company.

None of this was done because Texaco had been convicted of discrimination; it was done because Texaco had been *stigmatized*. And once stigmatized, the company plunged headlong into a series of reforms that allowed it to iconographically signal its racial virtuousness. Texaco did not reform so much as build an iconography of its own virtuousness to fend off stigma and restore moral authority. And, not surprisingly, its device of choice in this was that perfect icon, the racial preference. Goals and timetables, managerial evaluations based on the promotion of blacks, an oversight review board of outsiders—one gets the feeling that blacks are not as much preferred as they are hunted. How could a black take a job at Texaco without literally becoming an icon of the company's virtue, a piece of their redemption strategy? What manager would be foolish enough to fire a black for incompetence when he could promote him or her for profit?

This suggests that Texaco is not reforming; it is only stigma fighting. Racial discrimination was the original charge against the company, yet there is nothing in this blitz of iconographic virtue to suggest that the company has even learned what discrimination is. Had Texaco let the original discrimination suits against it go to trial, it might have learned exactly *how* it had discriminated (assuming it actually did). It could have then educated itself and the rest of America on this greatest of all social sins. Instead, what Texaco has shown America is that racial preferences are an amoral terrain where one can quickly buy immunity from the terror of stigma.

On the other hand, what a glorious victory the Texaco incident was for the grievance elite—an elite that thrives by turning grievance into entitlement, that lives by its appearance of wielding great power with the former oppressor. To walk in the front door with nothing but a gripe in one's pocket, and come back out in a matter of days with $176 million in hard cash—well, it would be very difficult to convince the crowd at the barbershop that this was not power. Again, patronage works as much by reputation as anything, so that even the black who will never see a dime of these millions has got to feel loyal on some level to the power that brought home so much cash in the name of his race. And not only cash, but also the *magic* of preferences that will instantly make equality where there was none before. Against the glamour of this power the idea of developmental reform grounded in principle is a hard sell.

And yet this is really a fool's power. It is only the power to pressure a stigmatized institution into erecting an iconography to better protect itself from stigma. At the end of the day the grievance elite has only the power to certify whatever iconographic reforms the institution comes up with. There may be jobs and preferences for some, but the reform itself is primarily iconographic and, therefore, preoccupied more with appearances than with the true development of the group. As dramatic a case as Texaco was, $176 million was not so much relative to the actual wealth of the company. Elite universities may have aggressive preference programs, but black students still have the highest dropout rate and the lowest grade-point average of any student group in America—*by far.* Six years after admissions, white students at the University of Colorado have a 72 percent graduation rate. The number for blacks is 39 percent. And wouldn't this 39 percent have been admitted and graduated *without* a preference? The preference often amounts to no more than an opportunity to fail at a higher level. So is the *power* that wins this sort of reform meaningful?

It is tragic that black America has had a leadership for the last thirty years that refuses to lead, that refuses to ask for meaningful development from its own people *whether or not "whites are kept on the hook"*—this last qualification being the measure of true leadership. The grievance elite, supported by spoils from the larger society, has chosen instead to pursue a power that, in essence, is merely emblematic.

3.

The loss of moral authority for whites and for American institutions that inevitably followed the early sixties civil rights victories transformed race into a problem that America simply had no *legitimate* power over. This opened a large vacuum in authority not only around race but around *all* areas of American life where social inequality might be an issue. And in the place of a legitimate power grounded in America's best principles, the power to stigmatize (whites with racism and blacks with uncle Tomism) became a new authority in its own right. Even today, in this territory the stigmatized are evil, the nonstigmatized are virtuous, and social morality is thus constituted. In the void of moral authority around race, stigma takes the place of principle by using shame as muscle.

This social territory where stigmatization rules, where atavisms find new legitimacy, where reform is only iconographic and merely deferential, and where atavistic preferences are imbued with magic—this territory might well be called *the culture of preference.*

In this culture, all flows from the threat of stigmatization. Here social virtue is *not accountable* to principle, honest struggle, or achievement, but only to iconography—to how virtuous its reforms *look.* Thus, the culture of preference is guided by an *ersatz* social virtue—a poor imitation of the real thing. Ersatz virtue is not simply a weak version of the real thing; it is not the real thing *at all.* Wherever the threat of stigmatization is immediate and powerful, ersatz virtue will thrive precisely to protect us from the risk of being stigmatized that real virtue always entails. And in the culture of preference social virtuousness comes without risk.

But if this culture is marked by ersatz virtue, it is even more visibly marked by a quality that often accompanies the ersatz: kitsch. We say that an artifact or a vision of life is kitsch when it is popular *because* it is rendered without reference to life's darkness and complexity. Kitsch is really a form of denial in which a glib and charming surface is created by repressing the bad and ugly elements of life. Milan Kundera gives it a more powerful, if far blunter, definition—". . . a world in which shit is denied and everyone acts as though it did not exist."

A UNICEF holiday card picturing children from many cultures standing together in warm harmony is kitsch because it represents a view of life shorn of the enmity, prejudice, and even war that cultural differences routinely inspire in the world. Because this picture represents a worldview that is fundamentally impossible, it is, despite all its easy appeal and charm, not really true.

The corruption of kitsch is that it invites *us* to feel that we are better than reality. By identifying with the UNICEF card we can entertain the idea that we are *essentially* as innocent and free of human evil as the image on the card. The idea is that this image represents the truth that at heart we, too, are without darkness and menace, and that it is only the corruptions of the world that sometimes involve us in dark impulses. We buy the card on some level to express this feeling about ourselves, this sense almost of a lost self that is better than the ruined world in which we live. Thus kitsch is always an invitation to a consoling sense of superiority.

All the ideas that define the culture of preference are kitsch.

"Diversity" is a term conceived to serve as kitsch. No two

people define it the same way. Both segregation and integration could be defended as forms of racial diversity. Like all kitsch, it is a word clean of all that is bad or difficult or complex. So it is a word that is useful for the very specific thing that it does *not* mean, which is racial prejudice. People use it to signal that they are *not* racially prejudiced. In other words, it is a word with no meaning whatsoever except that it *dissociates* people from the stigma of racism. "Inclusion," "pluralism," and even "multiculturalism" all work in the same way. Their truest purpose is to offer people a language through which they can console themselves with the feeling that they are superior to the nation's racial shame. This is the language of the culture of preference— a language that throws kitsch into war with stigma.

But this language of racial kitsch is hardly innocent. Like any other political kitsch, it is a corruption of power. It is a manipulation that makes the use of an undemocratic power seem no more than an innocent necessity. Kitsch allows the university admissions officer to say that he is not using a preference that racially discriminates against poor and hardworking Asians and whites in favor of better-off and underperforming blacks; rather he is pursuing diversity. Society has never voted to give him the power to discriminate racially, nor would the university administration go to its board of trustees and ask for the power to discriminate against Asians and whites. The 1964 Civil Rights Bill expressly prohibits exactly this kind of racial discrimination. Yet, because the kitsch of diversity answers the stigma of racism, he can tell himself that he is acting against racism even as he discriminates against Asians and whites *solely* on the basis of their race. He practices racism to escape the stigma of racism.

After all, diversity has no darkness or evil. Thus people who serve it can do no wrong.

Every demagogue tries to be a master of kitsch because it can legitimize otherwise blatantly illegitimate uses of power. "Reeducation" was the kitsch by which millions in the Communist world disappeared into gulags. A new "pacification program" was the way President Nixon extended a war the United States had turned against. In his support of preferential affirmative action, President Clinton asked an opponent of preferences if she would be against "the kind of affirmative action that produced Colin Powell." Of course, he never said what kind that was, whether it used preferences or not. Here General Powell himself became no more than a kitsch, an image like diversity that had no meaning except that it cut against the racist stigma. So the president was not supporting racially discriminatory preferences; he was fighting against the kind of racism that would prevent more Colin Powells from appearing.

The pernicious power of kitsch is that it draws us into an identification with ourselves *as we would be* if we were redeemed from whatever shame the kitsch opposes. Kitsches seduce us narcissistically with perfect images of ourselves, utterly free of the stigma to which we feel so vulnerable. For example, to identify with diversity is to see ourselves, as if in a mirror, cleansed of racial stigma. But the instant we make this identification, and use the kitsch for our own sense of virtuousness, we also give license to some agent of diversity. And, whether politician or admissions officer, this person will unilaterally seize the power to act in the name of diversity, as if from an unquestioned authority, and as if his or her action itself were the purest expression of racial innocence. And if this person discriminates . . . well, now in thrall to an image of our own perfect racial innocence, we are less inclined to see his or her actions as an abuse of power.

Political correctness is essentially a demagoguery of kitsch. It is a series of empty and banal words and ideas that iconographically oppose the racist and sexist stigma and, thus, license people to circumvent the normal avenues to power. Under a kitschy word like "inclusion" you can hire only women for two years, or set up an executive lounge only for blacks, or reject all Asians after a certain cutoff, or lower the bar three hundred SAT points for blacks, or set aside 30 percent of city contracts for minorities, and so on. Because the kitsch of "inclusion" has no evil, actions in its name can be only good.

In the culture of preference, kitsch is the primary source of power and license because it offers an imagery of racial innocence with which people and institutions can fend off the accusations of stigma.

I think the culture of preference grew out of that wrestling with stigma that was evident in William Styron and in the white friend I mentioned at the beginning of this essay, when they both protested that they felt no racial guilt. Even guilt-lessness brings no immunity to stigma. And the civil rights movement brought more than civil rights legislation: It also brought to blacks the power to stigmatize the race that had so long stigmatized them. My friend and Styron didn't like it, as well they shouldn't, but this new two-way street of stigmatization is what triggered the culture of preference.

The United States has constructed two enormous racial worlds in its history. The first, slavery and segregation, was built around the stigmatization of blacks as inferior, and the second, the culture of preference, was built to refute the stigmatization of whites as racists. Both have been essentially atavistic and antidemocratic worlds in which race has been legitimized as a criterion for discrimination and as a source of entitlement in clear defiance of every democratic principle. No doubt the second world is easier on everyone than the first, but it is not fundamentally different.

Though the second world was designed to redeem the shame of the first, I believe it has only given a pseudolegitimacy to the same racialism that shamed that first world. The culture of preference mimics the racial contradictions, discriminations, and the flaunting of democratic principle that cost America its moral authority around race to begin with. As a result, it has utterly failed to restore moral authority, which was its one and only mission. The duplicity of preferential treatment, the ineffectiveness of deferential reforms that only offer license while asking nothing of the people they seek to help, the banality and doublespeak of political correctness, the identity politics—all this and more has progressively alienated most Americans (never more than tolerant) from the culture of preference. The fate of this culture hangs on affirmative action because it legitimizes the violation of principles—allowing an atavism like color to be a source of entitlement—that supports all else in the culture. The anti-affirmative-action movement is also the movement against the entire culture of preference. It has won at least one popular vote—Proposition 209, which ends group preferences in California state government.

And yet the stigmatization of whites as racist remains so powerful that it keeps this culture alive well past the time when it has clearly lost respect (and despite the fact that it has never enjoyed wide support). One of the greatest failures of the culture of preference is that, for all its deference and preference, it was unable to lift the stigma from whites and American institutions. Corporate America is more vulnerable to stigma than ever before. The Republican Party avoids more than approaches legislation to end group preferences. And, more important, people across the country who truly disapprove of preferences simply remain silent. One wonders if Proposition 209 would have passed if people had been required to publicly declare their vote. It very likely passed not because Californians are racist, but because they genuinely dislike group preferences and were given the opportunity to vote that sentiment *without threat of stigma*. Many who dislike preferences, in and out of public life, look hopefully to the Supreme Court not only because it can declare them unconstitutional but because, in doing so, it can spare everyone from stigmatization.

If anything, the culture of preference has only strengthened the power of stigma in American life. When a society's government and institutions spend more than thirty years setting their best principles aside out of fear of stigmatization, they don't earn their way out from under stigma; they validate and strengthen stigmatization as a power. Thirty years ago when there was clearly less equality in American life for blacks, no black leader could walk into a corporation one day and come out the next with $176 million. Today Jesse Jackson can not only do that, but, on coming out, he can declare the deal no more than a "beginning."

As the culture of preference has lost face in the United States, it has simultaneously increased the power of stigma. As whites more and more see the policies of this culture as absurd, they are also *more* vulnerable to the stigmatization that spawned them. This, I believe, is what defines the great polarization that is so evident today between black and white Americans. Whites grow more fatigued with preferences, deference, and the power of stigma, while blacks become more invested in stigmatization as their only significant power in American life. And behind the power of stigma, giving it credibility, is the charge that America is still a racist society. Thus, by this alignment blacks are far more likely to believe that racism is a barrier to their aspirations than whites, who clearly benefit by asserting the opposite. This has given rise in black America to something like a belief in racism almost as a theological truth. For blacks, racism is power. For whites it is vulnerability.

The culture of preference has worsened all this because it has always responded to stigma rather than to principle. The more one gives in to stigma, the more it rules. It defines the terms of reform—deference, license, iconography. And then society finds itself debating over icons of racial goodwill rather than over problems. And all the while stigma grows stronger as a power, generating more and more iconography. The United States has simply become habituated to fighting stigma with iconography. When someone says, "I'd be against preferences except I don't know what we'd put in their place," he or she is really saying that without preferences "I would have no iconography with which to fight off stigma. I would have nothing to separate myself from the evil the stigma accuses me of, so it would look to all the world that I was synonymous with it. I would simply be seen as a racist."

It was a discipline of principle that finally saved the United States from that first racial world, which was erected around slavery and segregation. The true American identity, I believe, is simply the profoundness of our relationship to democratic principles *despite* our many failings and duplicities. It is as though every catastrophe of principle only drives the point home more deeply that these principles are all we have. Of course they offer no shelter from stigma. But then, the culture of preference with its elaborate iconography doesn't either. The United States has to accept its past as proof of its need for principles today. We all *know* what is right and fair and democratic. No stigma should make us afraid of this.

LIBERAL BIAS AND
THE ZONE OF DECENCY

1.

A local newspaper recently reported that the Boalt Hall law school at UC Berkeley was having trouble finding ways to "achieve diversity" now that racial preferences were no longer legal in the state of California. I have spoken to the reporter who wrote the story enough to know that she considers herself both an objective journalist and probably a political centrist. No doubt she is sensitive to the charge that there is a liberal bias in the media. Yet her report that the law school was looking for ways to "achieve diversity" reflected a conspicuous liberal bias. This school was not looking for "diversity"; it was searching for ways to bring in black and brown faces now that racial *preferences* (not affirmative action) were against the law. When the reporter used the word "diversity" without irony, she made an ideological concession to liberalism that biased and distorted the entire story.

Why would she do this?

Since 89 percent of the working press count themselves Democratic, it would be easy to assume that she intentionally let a little of her politics bleed into the story. But I don't think so. I think she was simply being polite.

Somewhere behind her use of this language was an idea of

decency, a sense of racial propriety, that led her to feel that it was better to say the law school was looking for "diversity" than to say that it was looking for blacks and Hispanics. So, in a sense, it would be unfair to accuse this reporter of an intentional bias. She was reporting on a racial matter, and in this arena a liberal bias is simple good manners.

2.

White Americans became a stigmatized group after the civil rights victories of the sixties. They became identified with the shame of white racism that the nation had finally acknowledged, and they fell under a kind of suspicion that amounted to a stigma. This added a new burden to white life in America: Since the sixties whites have had to *prove a negative*—that they are *not* racist—in order to establish their human decency where race is concerned. In the public realm, where they are most accountable for this decency and where there is risk of being stigmatized as racists, whites and American institutions must have a shorthand, some way of easily proving the negative, of gesturing that they are *not* racist. The bane of belonging to a stigmatized group is this *necessity* of having to prove a negative simply to gain even ground with others.

This necessity for whites, I believe, gave America what I have called redemptive liberalism—that peculiar, post-sixties, deferential liberalism that has been more interested in redeeming the moral authority of whites than in the mundane work of *earning* racial equality through a call for black development. But it was a main theme of this liberalism—deference to former victims of America's racial shame—that hardened into a shorthand, a set of manners that whites could use to show themselves free of racism.

And this is what leads to situations like the above, in which the reporter's adherence to good racial manners involved an assertion of liberal politics. One of the great victories of post-sixties liberalism was its evolution from a mere politics into a propriety. This has had the effect of making many of its key ideas ideologically invisible. For many Americans "diversity" is not an ideological term that justifies racial engineering and that even licenses institutions to discriminate racially; it is a commonplace notion of racial virtue. When a politics becomes a propriety, its mere arguments become conventional truth. As post-sixties liberalism has infused and redefined one American institution after another, it has done so as a protocol, an etiquette for white decency.

The pretense this protocol asks of whites is that they appear as though they have carved out of themselves precisely that underside of human nature that led to America's racial shame in the first place—the human capacity for projection and prejudice. Thus the correct white is not a wise person who accepts his potential for bigotry as a way of guarding against it; he is instead a circumscribed human being who pursues innocence through an impossible self-repression.

3.

But this liberal propriety also requires another kind of repression. A central theme of post-sixties liberalism is deference to blacks and minorities in the form of a grant of license—lowered college admission standards, welfare without a requirement of work or education, lowered educational expectations in the public schools, protection from open competition with others in many areas of employment and contracting, and so on. What makes this deference possible is a specific act of imagination on the part of whites: They must always imagine blacks *outside* the framework of individual responsibility.

Deference for whites is achieved through a discipline of the imagination by which whites forbid themselves the right (because of their lack of moral authority) to see a lack of responsibility as even a partial cause of black problems or a seizing of responsibility as even a partial solution to those problems. The deferential white must see blacks as *aresponsible.*

As the idea that makes white deference possible, black aresponsibility is the cornerstone of the etiquette that whites must follow in order to prove the negative. To exhibit their racial innocence, whites and American institutions must treat individual responsibility as an *impropriety* in relation to blacks. If they mention it as either a cause of or a solution to some black difficulty they not only breach propriety but also invoke the racist stigma. So it is not surprising to hear black problems discussed as though they were entirely the responsibility of whites. For example, from the article mentioned above one would think that the challenge of getting more blacks and Hispanics into Boalt Hall law school was solely the responsibility of the law school. By the etiquette of white decency, it would be a little disgraceful for the reporter to ask minority leaders if black and Hispanic prelaw students might assume more responsibility in making themselves competitive. Here white decency functions as ideology.

4.

But this idea of white decency also carries a very unfortunate implication: To imagine blacks as aresponsible is also to imagine them as inherently inferior. Black inferiority is an element in the discipline by which whites imagine black aresponsibility as a means to their own decency. This may be an unintended consequence, but it is a consequence nonetheless. To count blacks as aresponsible, they have to be seen as incapable of responsibility. Whether this incapacity comes from oppression or, as racists believe, from genetics, white decency requires that it be seen as intractable, something that will have to be permanently accounted for.

This is how the liberal bias in the media came to carry a clear implication of intractable black inferiority.

5.

In a recent *New Yorker* article on California after the elimination of race and gender preferences, the journalist Jeffrey Rosen argued that a group preference for blacks and Hispanics was the only way institutions could maintain standards for whites and Asians. His reasoning was simply that blacks and Hispanics would not be competitive on either standardized tests or grade-point averages. Without preferences to help them, there would be great pressure to lower institutional standards for everyone so that these groups could get in without truly competitive scores and grades. This, he rightly argued, would destroy elite state universities for everyone. He mentioned the famous case of New York's City College, which collapsed into mediocrity after decades of greatness because an open-enrollment policy shattered its standards. Thus he concludes that we should have low, preferential standards for blacks and Hispanics precisely so we can maintain high, selective standards for whites and Asians.

It is impossible not to notice the way this argument would reinforce the very racial hierarchy that has historically kept blacks at the bottom of society. And it is amazing to watch Rosen, in an article on affirmative action, leap over the challenge of pursuing racial equality through black development and, instead, concern himself with how to maintain meaningful standards for whites. He likes preferences because they give up on black excellence in order to preserve white excellence. They are a palliative to the hopelessly inferior that preserves the greatest white advantage of them all: high standards.

Rosen would surely say that keeping whites above blacks was not his true goal. However, it is the unintended result of his true goal, which is to establish his decency as a white by offering blacks the deference of preferences. He is operating out of the liberal ethic that makes it an impropriety to consider black difficulties within a framework of individual responsibility. Any demand that blacks take responsibility for meeting the same standards as others would be indecent—this despite the fact that most blacks applying to California's elite law schools are not disadvantaged. In the rush to preserve his own decency and high standards for even poor whites, he asks nothing of even well-off blacks. He exercises that peculiar liberal discipline of circumscription in which *all* blacks are imagined to be aresponsible and, by implication, intractably inferior.

It is a mistake to presume that the liberal bias is always in favor of blacks since it is usually based on an inference of their inferiority.

6.

"We believe the promise of America is equal opportunity, not equal outcomes." This statement is taken from the declaration of principles of the Democratic Leadership Conference in 1990, the year in which Governor Bill Clinton served as its chairman. It was five years later, in the summer of 1995, that President Bill Clinton came to the defense of precisely the kind of preferential affirmative action that would engineer outcomes in his famous "mend it; don't end it" speech. From the governor who had come to disbelieve in preferences to the president who would defend them. Why the change of heart?

I think one reason is that the preference is the perfect mechanism of deference. It *enforces* black aresponsibility. It not only implies that blacks should not be fully responsible, but it *prevents* them from assuming such responsibility. And if the preference in this way implies black inferiority, it also protects blacks from this inferiority by giving them a handicap in their competition with others.

But the president would not likely have taken note of all these mechanics. Like whites in general and the leadership of most American institutions, he would simply have noticed that his support of preferences lifted him beyond any association with America's racial shame and into what might be called a zone of decency.

The power whites gain by supporting preferences that deferentially lift responsibility from blacks gains them entrance into this zone of decency, where they become (1) immune to stigmatization as racists, (2) certified as racially decent, and (3) morally powerful enough to accept or reject difficult and expensive racial reforms with impunity. After President Clinton's "mend it; don't end it" speech and the announcement of his utterly symbolic "race dialogue," he went straight to the NAACP's 1997 national convention and told black leaders not to expect new social programs from the government.

However, the greatest power that whites and American institutions gain when they enter the zone of decency is the power to wield the same stigma that they escaped when they announced their support of preferences. In other words, once certified themselves, they gain the power to decertify others. In what I have called the culture of preference, a failure to support preferences indicates an interlocking series of moral failures that finally make one ineligible for decency. Not to support them is to fail to see blacks as aresponsible, which is to fail to offer them deference, which is to be indecent, uncertified, and associated with America's racial shame. The zone of decency makes the matter of racial morality virtually an either/or game, with everything turning on support or non-support of preferences. So, with a mere nodding support of preferential affirmative action one, in effect, stigmatizes all who don't support it.

Entering this zone not only shows that one has rejected America's shameful past; it also suggests that one has joined a new and higher American racial destiny. The zone is, in fact, a place of racial hope. Even its language—"multiculturalism," "inclusion," "diversity,"—seems to promise a grand racial future. And this idea of a beautiful if unspecified racial destiny redounds to enhance those within the zone even more: Added to their virtue is a link to destiny. Moreover, every degree of enhancement they enjoy adds a degree of stigmatization to those resisters of preferences who dwell outside the zone.

President Clinton must have discovered some of this in the years between his seeming rejection of preferences in 1990 and his new endorsement of them in his 1995 speech. In fact, it could be said that he made quite a find in preferences.

7.

Because the zone of decency claims the *entire* ground of racial virtue for itself, so that all outside its boundaries are uncertified as racially virtuous, it entitles those inside the zone to have power over social issues by default. The capacity of this zone to create a default that delivers power and entitlement to its insiders, while denying them to outsiders, is what makes the zone an enormous source of political power. If a politician can lift himself and his politics into this zone, he can create a moral default in his opposition so that it seems unworthy of moral leadership. By default he becomes worthy of such leadership because his politics seem linked to social virtue and destiny. He is certified as decent and is immune to racial stigma, and this is something people want in a political leader. In itself, it makes him attractive and modern.

Of course, no politician in modern times has mined this vein of power more effectively than President Clinton. Racial preferences for him are part of a larger fabric of political power. But it is important to remember that they don't enhance him much directly. Many of his supporters have very ambivalent feelings about them. Preferences enhance him indirectly by *decertifying* his opposition, by creating a moral default in them that redounds to him as moral worthiness in social affairs. He can talk to minorities, even tell them no in many instances, because of immunity purchased by preferences.

The zone of decency has been and remains a great source of liberal power in America, a source that has given liberalism a political and cultural power far beyond that warranted by the number of its followers. Though more than 80 percent of Americans (including large percentages from all ethnic groups) regularly poll against race and gender preferences, virtually all American institutions—public and private—have some form of preferential affirmative action. Amorphous liberal ideas like multiculturalism, of which most Americans are innately suspicious, have swept through the educational world altering the curriculum in arbitrary ways and justifying the development of ethnic fiefdoms. Diversity is currently sweeping through the corporate world in what is often an obvious shakedown of companies for preferences and money.

But, whether in a university or a corporation or the body politic itself, the zone of decency creates a moral default that decertifies the majority so that it has no moral authority upon which to speak out. So the ideas of liberalism have prevailed in America—especially within institutions where moral default can be easily created—as a minority ideology.

Conversely, American conservatism has an influence far smaller than its number of followers because its failure to support preferences keeps it in moral default. Without a support of preferences, conservatism has no way to fend off its rather severe stigmatization as a racially questionable ideology. The problem for conservatism, given its commitment to individual rights over group rights and to individual responsibility, is that it simply has no way to offer deference to blacks in the name of America's racial shame. It offers responsibility precisely where liberalism offers aresponsibility, so it always seems antagonistic where its competition is deferential. It has no zone of decency, no grand language of racial virtue and high destiny, no way to grant itself power by creating moral default in its opposition.

The real problem for conservatism is that post-sixties liberalism draws great moral authority by claiming to address America's historical racial shame. In many ways this liberalism feeds off that shame. It would never have its moral authority, and thus its disproportionate power, if it did not "work" this shame to some extent. But liberalism succeeds in this strategy because conservatism has failed to articulate how its principles and faith in the individual—its very disregard of race—in fact address this shame. After all, *reality is conservative.* And the timeless values that liberalism ceded to conservatism after the sixties—excellence, sacrifice, entrepreneurialism, discipline, and so on—have lifted far more minorities into the middle and even upper classes than the preferential and "aresponsible" ministrations of post-sixties liberalism. The moral authority of conservatism will come from the freedom from racism its insistence on principles can afford and the effectiveness of its values against the hard realities that so many minorities face. Today it is the illusion that reality is not conservative—fostered in the interest of liberal power—that defers the dream of black freedom.

American institutions paint themselves in the ideas of liberalism because they have been weakened by shame. The liberal bias in the media, corporate America, academia, government, and other institutions is an attempt to construct a white and institutional decency that will fend off this shame. But so far this decency has only deferred to shame and used its stigma politically. For America to survive its struggle with racial shame, it will have to stop cringing before it. This only makes for another kind of indecency. We should just bear this shame as a kind of wisdom, and earn a less glamorous decency by working strictly within the rules of democracy.

THE NEW SOVEREIGNTY

In the late 1960s, without much public debate but with many good intentions, the United States embarked on one of the most dramatic social experiments in its history. The federal government, radically and officially, began to alter and expand the concept of entitlement in America. Rights to justice and to government benefits were henceforth to be extended not simply to individuals but also to racial, ethnic, and other groups. Moreover, the essential basis of all entitlement in America—the guarantees of the Constitution—had apparently been found wanting; there was to be redress and reparation of past grievances, and the Constitution had nothing to say about that.

Martin Luther King, Jr., and the early civil rights leaders had demanded only constitutional rights; these had been found wanting too. By the late sixties, among a new set of black leaders, there had developed a presumption of collective entitlement (based on the redress of past grievances) that made blacks eligible for rights beyond those provided for in the Constitution, and thus beyond those afforded the nation's nonblack citizens. Thanks to the civil rights movement, a young black citizen as well as a young white citizen could not be turned away from a college because of the color of his or

her skin; by the early seventies a young black citizen, poor or
wealthy, now qualified for certain grants and scholarships—
might even be accepted for admission—simply *because* of the
color of his or her skin. This new and rather unexamined
principle of collective entitlement led America to pursue a
democracy of groups as well as of individuals—that collective
entitlement enfranchised groups just as the Constitution
enfranchised individuals.

I mentioned all this in a talk, "The New Sovereignty," I gave to
a university audience. In America today, I said, sovereignty—
that is, the power to act autonomously—is bestowed on any
group that is able to construct itself around a perceived grievance. With the concept of collective entitlement now accepted
not only at the federal level but casually at all levels of society,
any aggrieved group—and, for that matter, any assemblage of
citizens that might or might not previously have been thought
of as such a group—could make its case, attract attention and
funding, and build a constituency that, in turn, would increase
attention and funding. Soon this organized group of aggrieved
citizens would achieve sovereignty, functioning within our
long-sovereign nation and negotiating with that nation for a
separate, exclusive set of entitlements. And here I pointed to
America's university campuses, where, in the name of their
grievances, blacks, women, Hispanics, Asians, Native Americans, and gays and lesbians had hardened into sovereign constituencies that vied for the entitlements of sovereignty—separate "studies" departments for each group, "ethnic" theme
dorms, preferential admissions and financial aid policies, a
proportionate number of faculty of their own group, separate
student lounges and campus centers, and so on. This push for
equality among groups, I said, necessarily made for an inequality among individuals that prepared the ground for precisely

the racial, gender, and ethnic divisiveness that, back in the sixties, we all said we wanted to move beyond.

At the reception that followed the talk I was approached by a tall, elegant woman who introduced herself as the chairperson of the Women's Studies Department. Anger and the will to be polite were at war in her face, so that her courteous smile at times became a leer. She wanted to "inform" me that she was proud of the fact that Women's Studies was a separate department at her university. I asked her what could be studied in this department that could not be studied in other departments. Take the case of, say, Virginia Woolf: In what way would a female academic teaching in the Women's Studies Department have a different approach to Woolf's writing than a woman professor in the English Department? Above her determined smile her eyes became fierce. "You must know as a black that they won't accept us"—meaning women, blacks, presumably others—"in the English Department. It's an oppressive environment for women scholars. We're not taken seriously there." I asked her if that wasn't all the more reason to be there, to fight the good fight, and to work to have the contributions of women broaden the entire discipline of literary studies. She said I was naive. I said her strategy left the oppressiveness she talked about unchallenged. She said it was a waste of valuable energy to spend time fighting "old white males." I said that if women were oppressed, there was nothing to do *but* fight.

We each held tiny paper plates with celery sticks and little bricks of cheese, and I'm sure much body language was subdued by the tea-party postures these plates imposed on us. But her last word was not actually a word, it was a look. She parodied an epiphany of disappointment in herself, as if she'd caught herself in a bizarre foolishness. *Of course this guy is the enemy. He is the very oppressiveness I'm talking about. How could I have missed it?* And so, suddenly comfortable in the understanding that I was hopeless, she let her smile become gracious. Grace was something she could afford now. An excuse was made, a hand extended, and then she was gone. Holding my little plate, I watched her disappear into the crowd.

Today there are more than five hundred separate Women's Studies Departments and programs in American colleges and universities. There are nearly four hundred independent Black Studies Departments or programs, and hundreds of Hispanic, Asian, and Native American programs. Given this degree of entrenchment, it is no wonder this woman found our little debate a waste of her time. She would have had urgent administrative tasks awaiting her attention—grant proposals to write, budget requests to work up, personnel matters to attend to. And suppose I had won the debate? Would she have rushed back to her office and begun to dismantle the Women's Studies Department by doling out its courses and faculty to long-standing departments like English and History? Would she have given her secretary notice and relinquished her office equipment? I don't think so.

I do think I know how it all came to this—how what began as an attempt to address the very real grievances of women wound up creating newly sovereign fiefdoms like this Women's Studies Department. First there was collective entitlement to redress the grievances, which in turn implied a sovereignty for the grievance group, since sovereignty is only the formalization of collective entitlement. Then, since sovereignty requires autonomy, there had to be a demand for separate and independent stature within the university (or some other institution of the society). There would have to be a separate territory, with the trappings that certify sovereignty and are concrete recognition of the grievance identity—a building or suite of offices, a budget, faculty, staff, office supplies, letterhead, and so on.

And so the justification for separate women's and ethnic studies programs has virtually nothing to do with strictly academic matters and everything to do with the kind of group-identity politics in which the principle of collective entitlement has resulted. My feeling is that there can be no full redress of the woeful neglect of women's intellectual contributions until those contributions are entirely integrated into the very departments that neglected them in the first place. The same is true for America's minorities. Only inclusion answers exclusion. But now all this has been confused by the sovereignty of grievance group identities.

It was the sovereignty issue that squelched my talk with the Women's Studies chairperson. She came to see me as an enemy not because I denied that women writers had been neglected historically; I was the enemy because my questions challenged the territorial sovereignty of her department and the larger grievance identity of women. It was not a matter of fairness—of justice—but of power. She would not put it that way, of course. For in order to rule over her sovereign fiefdom it remains important that she seem to represent the powerless, the aggrieved. It remains important, too, that my objection to the new sovereignty can be interpreted by her as sexist. When I failed to concede sovereignty, I became an enemy of women.

In our age of the New Sovereignty the original grievances—those having to do with fundamental questions such as basic rights—have in large measure been addressed, if not entirely redressed. But this is of little matter now. The sovereign fiefdoms are ends in themselves—providing career tracks and bases of power. This power tends to be used now mostly to defend and extend the fiefdom, often by exaggerating and exploiting secondary, amorphous, or largely symbolic complaints. In this way, the United States has increasingly become an uneasy federation of newly sovereign nations.

In *The True Believer*, Eric Hoffer wrote presciently of this phenomenon I have called the New Sovereignty: "When a mass movement begins to attract people who are interested in their individual careers, it is a sign that it has passed its vigorous stage; that it is no longer engaged in molding a new world but in possessing and preserving the present. It ceases then to be a movement and becomes an enterprise."

If it is true that great mass movements begin as spontaneous eruptions of long-smoldering discontent, it is also true that after significant reform is achieved they do not like to disappear or even modify their grievance posture. The redressing of the movement's grievances wins legitimacy for the movement. Reform, in this way, also means recognition for those who struggled for it. The movement's leaders are quoted in the papers, appear on TV, meet with elected officials, write books—they come to embody the movement. Over time they and they alone speak for the aggrieved; and of course they continue to speak *of* the aggrieved, adding fresh grievances to the original complaint. It is their vocation now, and their means to status and power. The idealistic reformers thus become professional spokespersons for the seemingly permanently aggrieved. In the civil rights movement, suits and briefcases replaced the sharecropper's denim of the early years, and five-hundred-dollar-a-plate fund-raisers for the National Association for the Advancement of Colored People (NAACP) replaced volunteers and picket signs. The raucous bra burning of late-sixties feminism gave way to Women's Studies Departments and direct-mail campaigns by the National Organization of Women (NOW).

This sort of evolution, however natural it may appear, is not without problems for the new grievance-group executive class. The winning of reform will have dissipated much of the explosive urgency that started the movement, yet the new institutionalized movement cannot justify its existence without this urgency. The problem becomes one of maintaining a reformist organization after considerable reforms have been won.

To keep alive the urgency needed to justify itself, the grievance organization will do three things. First, it will work to inspire a perpetual sense of grievance in its constituency so that grievance becomes the very centerpiece of the group itself. To be black, or a woman, or gay is, in the eyes of the NAACP, NOW, or Act Up, to be essentially threatened, victimized, apart from the rest of America. Second, these organizations will up the ante on what constitutes a grievance by making support of sovereignty itself the new test of grievance. If the Women's Studies Department has not been made autonomous, this constitutes a grievance. If the National Council of La Raza hasn't been consulted, Hispanics have been ignored. The third strategy of grievance organizations is to arrange their priorities in a way that will maximize their grievance profile. Often their agendas will be established more for their grievance potential than for the actual betterment of the group. Those points at which there is resistance in the larger society to the group's entitlement demands will usually be made into top-priority issues, thereby emphasizing the status of victim and outsider necessary to sustain the sovereign organization.

Thus, at its 1989 convention, the NAACP put affirmative action at the very top of its agenda. Never mind the fact that studies conducted by both proponents and opponents of affirmative action indicated that the practice has very little real impact on the employment and advancement of blacks. Never mind, too, that surveys show that most black Americans do not consider racial preferences *their* priority. In its wisdom the

NAACP thought (and continues to think) that the national mood against affirmative-action programs is a bigger problem for black men and women than is teen pregnancy, or the disintegrating black family, or black-on-black crime. Why? Because the very resistance affirmative action meets from the larger society makes it an issue of high grievance potential. Affirmative action can generate the urgency that justifies black sovereignty far more effectively than such issues as teen pregnancy or high dropout rates, which carry no load of collective entitlement and which the *entire* society sees as serious problems.

In the women's movement, too, the top-priority issues have been those with the highest grievance potential. I think so much effort and resources went into the now-failed Equal Rights Amendment (ERA) because, in large part, it carried a tremendous load of collective entitlement (a constitutional amendment for a specific group rather than for all citizens) and because it faced great resistance from the larger society. It was a win-win venture for the women's movement. If it succeeded there would be a great bounty of collective entitlement; if it failed, as it did, the failure could be embraced as a grievance—an indication of America's continuing unwillingness to assure equality for women. *America does not want to allow us in!*—that is how the defeat of ERA could be interpreted by NOW executives and by female English professors eager to run their own departments: The defeat of the ERA was a boon for the New Sovereignty.

I also believe that this quest for sovereignty at least partially explains the leap of abortion rights to the very top of the feminist agenda on the heels of the ERA's failure. Abortion has always been an extremely divisive, complex, emotionally charged issue. And for this reason it is also an issue of enormous grievance potential for the women's movement—assuming it can be framed solely in terms of female grievance. My own belief is that abortion is a valid and important issue for the women's movement to take up, and I support the pro-choice position the movement advocates. However, I think that women's organizations like NOW have framed the issue in territorial terms in order to maximize its grievance potential. When they make women's control of their own bodies the very centerpiece of their argument for choice, they are making the fact of pregnancy the *exclusive* terrain of women, despite the obvious role of men in conception and despite the fact that the vast majority of married women who decide to have abortions reach their decisions with their husbands. Framed exclusively as a woman's right, abortion becomes not a societal issue or even a family issue but a grievance issue in the ongoing struggle of the women's movement. Can women's organizations continue to frame pro-choice as a grievance issue—a question of a right—and expect to garner votes in Congress or in state legislatures, which is where the abortion question is headed?

I don't think this framing of the issue as a right is so much about abortion as it is about the sovereignty and permanency of women's organizations. The trick is exclusivity. If you can make the issue exclusively yours—within your territory of final authority—then all who do not capitulate are aggrieving you. And then, of course, you must rally and expand your organization to meet all this potential grievance.

But this is a pattern that ultimately puts grievance organizations out of touch with their presumed constituencies, who grow tired of the hyperbole. I think it partially explains why so many young women today resist the feminist label, and why the membership rolls of the NAACP have fallen so sharply in recent years, particularly among the young. The high grievance profile is being seen for what it mostly is—a staying-in-business strategy.

*   *   *

How did the United States evolve its now rather formalized notion that groups of its citizens could be entitled collectively? I think it goes back to the most fundamental contradiction in American life. From the beginning America has been a pluralistic society, and one drawn to a radical form of democracy—emphasizing the freedom and equality of individuals—that could meld such diversity into a coherent nation. In this new nation no group would lord it over any other. But, of course, beneath this America of its ideals there was from the start a much meaner reality, one whose very existence mocked the notion of a nation made singular by the equality of its individuals. By limiting democracy to their own kind—white, male landowners—the Founding Fathers collectively entitled themselves and banished all others to the edges and underside of American life. There individual entitlement was either curtailed or—in the case of slavery—extinguished.

The genius of the civil rights movement that changed the fabric of American life in the late fifties and early sixties was its profound understanding that the enemy of black America was not the ideal America but the unspoken principle of collective entitlement that had always given the lie to true democracy. This movement, which came to center stage from America's underside and margins, had as its single, overriding goal the eradication of white entitlement. And, correspondingly, it exhibited a belief in democratic principles at least as strong as that of the Founding Fathers, who themselves had emerged from the (less harsh) margins of English society. In this sense the civil rights movement reenacted the American Revolution, and its paramount leader, Martin Luther King, Jr., spoke as twentieth-century America's greatest democratic voice.

All this was made clear to me for the umpteenth time by my father on a very cold Saturday afternoon in 1959. There was a national campaign under way to integrate the lunch counters at Woolworth stores, and my father, who was more a persuader than an intimidator, had made it a point of honor that I join him on the picket line, civil rights being nothing less than a religion in our household. By this time, at age twelve or so, I was sick of it. I'd had enough of watching my parents heading off to still another meeting or march; I'd heard too many tedious discussions on everything from the philosophy of passive resistance to the symbolism of going to jail. Added to this, my own experience of picket lines and peace marches had impressed on me what so many people who have partaken of these activities know: that in themselves they are crushingly boring—around and around and around holding a sign, watching one's own feet fall, feeling the minutes like hours. All that Saturday morning I hid from my father and tried to convince myself of what I longed for—that he would get so busy that if he didn't forget the march he would at least forget me.

He forgot nothing. I did my time on the picket line, but not without building up enough resentment to start a fight on the way home. What was so important about integration? We had never even wanted to eat at Woolworth's. I told him the truth, that he never took us to *any* restaurants anyway, claiming always that they charged too much money for bad food. But he said calmly that he was proud of me for marching and that he knew *I* knew food wasn't the point.

My father—forty years a truck driver, with the urges of an intellectual—went on to use my little rebellion as the occasion for a discourse, in this case on the concept of integration. Integration had little to do with merely rubbing shoulders with white people, eating bad food beside them. It was about the right to go *absolutely* anywhere white people could go being the test of freedom and equality. To be anywhere they could be, do anything they could do, was the point. Like it or not, white people defined the horizon of freedom in America, if you couldn't touch their shoulder you weren't free. For him integration was *evidence* of freedom and equality.

My father was a product of America's margins, as were all the blacks in the early civil rights movement, leaders and foot soldiers alike. For them integration was a way of moving from the margins into the mainstream. Today there is considerable ambivalence about integration, but in that day it was nothing less than democracy itself. Integration is also certainly about racial harmony, but it is more fundamentally about the ultimate extension of democracy—beyond the racial entitlements that contradict it. The idea of racial integration is quite simply the most democratic principle America has evolved, since all other such principles depend on its reality and are diminished by its absence.

But the civil rights movement did not account for one thing: the tremendous release of black anger that would follow its victories. The 1964 Civil Rights Act and the 1965 Voting Rights Act were, on one level, admissions of guilt by American society that it had practiced white entitlement at the expense of all others. When the oppressors admit their crimes, the oppressed can give full vent to their long-repressed rage because now there is a moral consensus between oppressor and oppressed that a wrong was done. This consensus gave blacks the license to release a rage that was three centuries deep, a rage that is still today everywhere visible, a rage that—in the wake of the Rodney King verdict, a verdict a vast majority of Americans thought unfair—fueled the worst rioting in American history.

By the mid-sixties the democratic goal of integration was no longer enough to appease black anger. Suddenly for blacks there was a sense that far more was owed, that a huge bill was due. And for many whites there was also the feeling that some kind of repayment was truly in order. This was the moral logic that followed inevitably from the new consensus. But it led to an even simpler logic: If blacks had been oppressed collec-᾽ tively, that oppression would now be redressed by entitling them collectively. So here we were again, in the name of a thousand good intentions, falling away from the hard challenge of a democracy of individuals and embracing the principle of collective entitlement that had so corrupted the American ideal in the first place. Now this old sin would be applied in the name of uplift. And this made an easy sort of sense. If it was good enough for whites for three hundred years, why not let blacks have a little of it to get ahead? In the context of the sixties—black outrage and white guilt—a principle we had just decided was evil for whites was redefined as a social good for blacks. And once the formula was in place for blacks, it could be applied to other groups with similar grievances. By the seventies more than 60 percent of the American population—not only blacks but Hispanics, women, Asians—would come under the collective entitlement of affirmative action.

In the early days of the civil rights movement, the concept of solidarity was essentially a moral one. That is, all people who believed in human freedom, fairness, and equality were asked to form a solid front against white entitlement. But after the collaboration of black rage and white guilt made collective entitlement a social remedy, the nature of solidarity changed. It was no longer the rallying of diverse people to breach an oppressive group entitlement. It was the very opposite: a rallying of people within a grievance group to pursue their own group entitlement. As early as the mid-sixties, whites were made unwelcome in the civil rights movement, just as, by the mid-seventies, men were no longer welcome in the women's movement. Eventually collective entitlement *always* requires separatism. And the irony is obvious: Those who once had been the victims of separatism, who had sacrificed so dearly to overcome being at the margins, would later create an ethos of their own separatism. After the sixties solidarity became essentially a separatist concept, an exclusionary principle. One no longer heard words like "integration" or "harmony"; one heard about "anger" and "power." Integration is anathema to grievance groups for precisely the same reason it was anathema to racist whites in the civil rights era: because it threatens their collective entitlement by insisting that no group be entitled over another. Power is where it's at today—power to set up the organization, attract the following, run the fiefdom.

But it must also be said that this could not have come to pass without the cooperation of the society at large and its institutions. Why did the government, the public and private institutions, the corporations and foundations, end up supporting principles that had the effect of turning causes into sovereign fiefdoms? I think the answer is that those in charge of America's institutions saw the institutionalization and bureaucratization of the protest movements as ultimately desirable, at least in the short term, and the funding of group entitlements as ultimately a less costly way to redress grievances. The leaders of the newly sovereign fiefdoms were backing off from earlier demands that the United States live up to its ideals. Gone was the moral indictment. Gone was the call for difficult, soulful transformation. The language of entitlement is essentially the old, comforting language of power politics, and in the halls of power it went down easily enough.

With regard to civil rights, the moral voice of Dr. King gave way to the demands and cajolings of poverty-program moguls, class-action lawyers, and community organizers. The compromise that satisfied both political parties was to shift the focus from democracy, integration, and developmental uplift to collective entitlements. This satisfied the institutions because entitlements were cheaper in every way than real change. Better to set up Black Studies and Women's Studies Departments than to have wrenching debates within existing departments. Better to fund these new institutions clamoring for money because who knew what kind of fuss they'd make if their proposals were turned down. Better to pass laws permitting Hispanic students to get preferred treatment in college admission—it costs less than improving kindergartens in East Los Angeles.

And this way to uplift satisfied the grievance-group "experts" because it laid the ground for their sovereignty and permanency: You negotiated with *us*. You funded *us*. You shared power, at least a bit of it, with *us*.

This negotiation was carried out in a kind of quasi secrecy. Quotas, set-asides, and other entitlements were not debated in Congress or on the campaign trail. They were implemented by executive orders and Equal Employment Opportunity Commission (EEOC) guidelines without much public scrutiny. Also the courts played a quiet but persistent role in supporting these orders and guidelines and in further spelling out their application. Universities, corporations, and foundations implemented their own grievance entitlements, whose workings are often kept from the public.

Now, it should surprise no one that all this entitlement has most helped those who least need it—white middle-class women and the black and Hispanic middle class. Poor blacks do not guide the black grievance groups. Working-class women do not set NOW's agenda. Poor Hispanics do not clamor for bilingualism. Perhaps there is nothing wrong with middle-class people being helped, but their demands for entitlements are most often in the name of those less well off than themselves. The negotiations that settled on entitlements as the primary form of redress after the sixties have generated a legalistic grievance industry that argues the interstices of entitlements and does very little to help those truly in need.

In a liberal democracy, collective entitlements based on race, gender, ethnicity, or some other group grievance are always undemocratic expedients. Integration, on the other hand, is the most difficult and inexpedient expansion of the democratic ideal; for in opting for integration, a citizen denies his or her impulse to use our most arbitrary characteristics—race, ethnicity, gender, sexual preference—as the basis for identity, as a key to status, or for claims to entitlement. Integration is twentieth-century America's elaboration of democracy. It eliminates such things as race and gender as oppressive barriers to freedom, as democrats of an earlier epoch eliminated religion and property. Our mistake has been to think of integration only as a utopian vision of perfect racial harmony. I think it is better to see integration as the inclusion of all citizens into the same sphere of rights, the same range of opportunities and possibilities that our Founding Fathers themselves enjoyed. Integration is not social engineering or group entitlements; it is a fundamental *absence* of arbitrary barriers to freedom.

If we can understand integration as an absence of barriers that has the effect of integrating all citizens into the same sphere of rights, then it can serve as a principle of democratic conduct. Anything that pushes anybody out of this sphere is undemocratic and must be checked, no matter the good intentions that seem to justify it. Understood in this light, collective entitlements are as undemocratic as racial and gender discrimination, and a group grievance is no more a justification for entitlement than the notion of white supremacy was at an earlier time. We were wrong to think of democracy as a gift of freedom; it is really a discipline that avails freedom. Some-times its enemy is racism and sexism; other times the enemy is our expedient attempts to correct these ills.

I think it is time for those who seek identity and power through grievance groups to fashion identities apart from grievance, to grant themselves the widest range of freedom, and to assume responsibility for that freedom. Victimhood lasts only as long as it is accepted, and to exploit it for an empty sovereignty is to accept it. The New Sovereignty is ultimately a vanity. It is a narcissism of victims, and it brings only a negligible power at the exorbitant price of continued victimhood. And all the while integration remains the real work.

Made in the USA
Monee, IL
13 June 2020